EARTH*RISE*

Earth*rise*

How Man First Saw *the Earth*

Robert Poole

YALE UNIVERSITY PRESS
NEW HAVEN AND LONDON

For information about this and other Yale University Press publications, please contact:
U.S. Office: sales.press@yale.edu   www.yalebooks.com
Europe Office: sales @yaleup.co.uk   www.yaleup.co.uk

Set in Minion by IDSUK (DataConnection) Ltd
Printed in the United States of America

Library of Congress Cataloging-in-Publication Data

Poole, Robert
    Earthrise : how man first saw the Earth / Robert Poole.
       p. cm.
    Includes bibliographical references and index.
    ISBN 978-0-300-13766-8 (ci : alk. paper)
    1. Earth--Photographs from space. 2. Earth--Remote-sensing images.
    I. Title.
    QB637.P66 2008
    525.022'2--dc22

                                                                    2008026764

A catalogue record for this book is available from the British Library.
10 9 8 7 6 5 4 3 2 1

To Yuko

Contents

Illustrations

Acknowledgements

Historians who step outside their accustomed fields need a variety of encouragement and assistance, and for me this has come in England from Ian Burney, Penelope Corfield, Alan Farmer, David Galbraith, Jeff Hughes, Rob Iliffe, Cliff O'Neill and John Swift. For Yale, Dennis Cosgrove has been an exceptionally thorough and generous critic, and Heather McCallum and Rachael Lonsdale have sharpened the book up no end. In the US I have enjoyed both the help and the interest of Bill Larsen and Glen Swanson at the Johnson Space Center; Jane Odom, Colin Fries, John Hargenrader and Liz Suckow at the NASA History Office; Neil Spencer at the Keith T. Glennan Memorial Library at NASA HQ; Gwen Pitman at NASA Media Services; Melissa Keiser, Mark Taylor, Roger Launius and David Devorkin at the National Air and Space Museum; Patricia Gossel at the National Museum of American History; and Jennifer Huergo at the Applied Physics Laboratory, Johns Hopkins University. Frank Borman, James Lovell, Stewart Brand and John and Anna McConnell have kindly provided recollections and information, and Richard Underwood has generously shared at length the thoughts and memories of his remarkable life. None of these is responsible for the errors and misunderstandings that doubtless remain. A period as visiting research fellow in the Department of History, University of Manchester, enabled me to launch this project, and support from the University of Cumbria has been valuable in helping it along, particularly that of Hugh Cutler and Sonia Mason in the Research Office. The University of Manchester John Rylands Library has been an invaluable resource, along with the libraries of the Institute of Historical Research, the University of London, Lancaster University, and the University of Cumbria, as well as the excellent Lancashire County Libraries Service. Elements of this book were presented at seminars at the Centre for the History of Science, Technology and Medicine at Manchester University, Lancaster University and the University of Cumbria, and I am grateful to all involved. Yuko Poole has lived with this book for as long as she has lived with me, and it is dedicated to her.

Timeline

4th century BC	Plato's description of the whole Earth 'composed of colours more numerous and beautiful than we have ever seen.'
1st century BC–2nd century AD	Age of the Roman *somnium*, or visionary dream of Earth, as recounted by Cicero, Ovid, Seneca and Lucian.
1543	Copernicus displaces the Earth from the centre of the universe.
1630s	Europe's lunar moment: Kepler and others imagine journeying to the Moon and seeing the Earth.
1791	Volney's vision of the Earth in *Ruins of Empire*.
1870	Jules Verne's description of the Earth in *Around the Moon*.
1900	World's Fair, Paris, inspiring Albert Kahn's 'Archives of the Planet' project to photograph the world (*c.*1908–30)
1916	Konstantin Tsiolkovsky, *Beyond the Planet Earth* (begun 1896).
1926	Vladimir Vernadsky, *The Biosphere*.
1931	David Lasser, *The Conquest of Space*, predicts that the sight of the whole Earth will erode human divisions.
1935	Explorer II balloon takes first photograph of the curvature of the Earth (13 miles up).
1945	End of Second World War; United Nations founded.

1946	V-2 rockets begin to photograph the curve of the Earth (65 miles up).
1948	V-2 panorama of one-tenth of the curve of the Earth released.
	UN Universal Declaration of Human Rights.
1950	Fred Hoyle predicts that a photograph of the whole Earth will unleash 'a new idea as powerful as any in history.'
1957	Sputnik satellite (USSR); start of the first space age (1957–72).
1959	Atlas rocket (USA) photographs one-sixth of the Earth's circumference in colour (800 miles up). Explorer satellite (USA) transmits first crude image of the crescent Earth (17,000 miles up). Scientific search for extra-terrestrial intelligence begins.
1960	Tiros weather satellite begins transmitting pictures of Earth from orbit (450 miles up).
1961	First man in space, Yuri Gagarin (USSR).
1961–3	Manned Mercury programme (USA) yields photographic prints of Earth from orbit (*c.* 100 miles up).
1963	Nuclear Test Ban Treaty prohibits nuclear explosions in space.
1965–6	Manned Gemini programme (USA) yields first high-quality orbital pictures of the Earth (100–800 miles up).
1966	Barbara Ward's *Spaceship Earth* published. James Lovelock's revelation of the Earth as a living system.
	February: Stewart Brand's campaign for a picture of the whole Earth.
	August: unmanned Lunar Orbiter transmits first picture of Earth from the Moon (240,000 miles out).
	December: ATS-I weather satellite transmits good black-and-white pictures of the whole Earth (22,300 miles up).

1967	January: UN Outer Space Treaty signed.
	August: DODGE satellite transmits first artificial colour picture of Earth (21,000 miles up).
	November: unmanned Apollo 4 brings back colour photos of the crescent whole Earth (11,000 miles up).
	November: ATS-III satellite transmits first full colour picture of the whole Earth (22,300 miles up).
1968	April: film *2001: a Space Odyssey*.
	Fall: first *Whole Earth Catalog*.
	November: unmanned Zond 6 (USSR) brings back first black-and-white photographic print of Earthrise over the Moon.
	December: Apollo 8 mission photographs Earthrise in colour.
1969	Friends of the Earth founded.
	March: Russell Schweickart's 'no frames, no boundaries' spacewalk on Apollo 9.
	Summer: Buckminster Fuller's 'World Game' played by students across America and Europe.
	July: Apollo 11 Moon landing; Earth flag displayed among crowds watching in New York.
1970	March 21: John McConnell's Earth Day, San Francisco.
	April 22: Gaylord Nelson's Earth Day, eastern USA.
1972	*Limits to Growth* report argues that the Earth's economy has environmental limits.
	June: UN Conference on the Human Environment, Stockholm – the first Earth Summit.
	December: Apollo 17 'Blue marble' photograph.
1974–5	Lindisfarne conferences on 'Planetary Consciousness'.
1977	Voyager 1 photographs Earth and Moon together.

1979	James Lovelock, *Gaia: a New Look at Life on Earth*.
1982–3	'Beyond War' movement film *No Frames, No Boundaries*; Jonathan Schell's book *The Fate of the Earth*; Carl Sagan's 'Nuclear Winter' hypothesis.
1985	First Association of Space Explorers conference, 'The Planet, Our Home.'
1988	UN Intergovernmental Panel on Climate Change (IPCC) established.
1990	February: Voyager 'Pale Blue Dot' photograph (four billion miles out). April: Earth Day revived.
1992	Second UN Earth Summit, Rio de Janeiro.
1997	Kyoto Agreement on Global Warming.
1999	Landsat 7 begins continuous global environmental monitoring.
2007	Third UN Earth Summit, Bali. Al Gore and IPCC win Nobel Peace Prize. Japanese Kaguya probe transmits high-definition film of Earthrise from the Moon.

Abbreviations

APL	Applied Physics Laboratory, John Hopkins University
JSC	Johnson Space Center
LOC	Library of Congress
NASA	National Aeronautics and Space Agency
NASM	National Air and Space Museum, Smithsonian Institution
NHO	NASA History Office

CHAPTER 1

Earthrise, seen for the first time by human eyes

On Christmas Eve 1968 three American astronauts were in orbit around the Moon: Frank Borman, James Lovell and Bill Anders. The crew of Apollo 8 had been declared by the United Nations to be the 'envoys of mankind in outer space'; they were also its eyes.[1] They were already the first people to leave Earth's orbit, the first to set eyes on the whole Earth, and the first to see the dark side of the Moon, but the most powerful experience still awaited them. For three orbits they gazed down on the lunar surface through their capsule's tiny windows as they carried out the checks and observations prescribed for almost every minute of this tightly planned mission.

On the fourth orbit, as they began to emerge from the far side of the Moon, something happened. They were still out of radio contact with the Earth, but the onboard voice recorder captured their excitement.

> Borman: Oh my God! Look at that picture over there! Here's the Earth coming up. Wow, that is pretty!
> Anders: Hey, don't take that, it's not scheduled.
> Borman: (Laughter). You got a colour film, Jim?
> Anders: Hand me that roll of colour quick, will you –
> Lovell: Oh man, that's great!
> Anders: Hurry. Quick . . .
> Lovell: Take several of them! Here, give it to me . . .
> Borman: Calm down, Lovell.[2]

The crew of Apollo had seen the Earth rise. The commander, Frank Borman, later recalled the moment.

I happened to glance out of one of the still-clear windows just at the moment the Earth appeared over the lunar horizon. It was the most beautiful, heart-catching sight of my life, one that sent a torrent of nostalgia, of sheer homesickness, surging through me. It was the only thing in space that had any color to it. Everything else was either black or white, but not the Earth.[3]

'Raging nationalistic interests, famines, wars, pestilences don't show from that distance,' he commented afterwards. 'We are one hunk of ground, water, air, clouds, floating around in space. From out there it really is "one world."' 'Up there, it's a black-and-white world,' explained James Lovell. 'There's no color. In the whole universe, wherever we looked, the only bit of color was back on Earth. . . . It was the most beautiful thing there was to see in all the heavens. People down here don't realize what they have.'[4] Bill Anders recalled how the moment of Earthrise 'caught us hardened test pilots'.

We'd spent all our time on Earth training about how to study the Moon, how to go to the Moon; it was very lunar orientated. And yet when I looked up and saw the Earth coming up on this very stark, beat up lunar horizon, an Earth that was the only color that we could see, a very fragile looking Earth, a very delicate looking Earth, I was immediately almost overcome by the thought that here we came all this way to the Moon, and yet the most significant thing we're seeing is our own home planet, the Earth.

Looking back after twenty years, Anders told a reporter that although he now thought only occasionally about those events, 'it was that Earth that really stuck in my mind when I think of Apollo 8. It was a surprise; we didn't think about that.'[5]

Like the crew of Apollo 8, NASA was so preoccupied with the Moon that it too forgot about the Earth. Photographs of Earth hardly featured at all on the official mission plans; they belonged in a miscellaneous category labelled 'targets of opportunity' and given the lowest priority. As for the television camera that would provide the first live pictures of the Earth, the coverage was amateurish and ill-prepared: there was trouble with the telephoto lenses, the camera was hard to aim and the capsule windows were fogged. There had been no official planning about what the astronauts would say when they made the first broadcasts to their home planet from the Moon. Perhaps surprisingly, they did not talk (as senior NASA controllers had anticipated) about 'one world' and 'peace on Earth' at Christmas; they read the opening verses of the Bible, the Creation story from the Book of Genesis. This was the one part of the publicity that had been prepared in advance, but it had been done by the astronauts themselves.

This general lack of preparedness had one important effect on all concerned: the sight of Earth came with the force of a revelation, a sense which deepened as the excitement of Apollo faded. After watching the last Apollo launch, the New Age philosopher William Irwin Thompson wrote: 'the recovery of our lost cosmic orientation will probably prove to be more historically significant than the design of the Saturn V rocket.' The writer Norman Cousins told the 1975 Congressional hearings on the future of the space programme: 'what was most significant about the lunar voyage was not that men set foot on the Moon, but that they set eye on the Earth.'[6]

This was not what had been supposed to happen. The cutting edge of the future was to be in space; Earth was the launchpad, not the target. Generations of techno-prophets had quoted the Russian space visionary Konstantin Tsiolkovsky (1869–1923): 'the Earth is man's cradle, but one cannot live in the cradle forever.'[7] Thinking about space in the 1950s and 1960s was dominated by a core of progressive ideas that were so taken for granted that they have only recently been given a name: astrofuturism. The astrofuturist ideal was that humanity would be revitalised by discovering its true destiny in space. It was a concept that could be found spread across the political spectrum from left to right (with perhaps a slight tendency to sag in the middle), and it was propagated through both speculative science writing and science fiction. Some prominent figures achieved equal fame in both fields: Arthur C. Clarke, Isaac Asimov and Fred Hoyle. Generations of rocket pioneers and space scientists were inspired in their youth by science fiction, and some went on to write it, including the Nazi rocketeer Wernher von Braun and the astronomer Carl Sagan.[8]

Wernher von Braun, salvaged from Germany with his V-2s in 1945 to form the basis for the American space programme, proved an adept publicist for space travel. While shrewdly alarming the generals with the line that control of space meant control of the world, he worked happily with influential media organisations such as Disney and the publishers of *Collier's Magazine*, on a series of articles, books, TV documentaries and even a theme park, all designed to make space seem real. 'Read it today, live it tomorrow!' ran the slogan of one science fiction magazine. Clumsily authentic science fiction films such as *Destination Moon* and *The Conquest of Space*, coupled with a surge of both science fiction and popular science writing, reinforced the message that space travel was inevitable, tapping into a bedrock of postwar belief in the prospect of limitless human advancement through technology.[9] 'Man has already poked his nose into space and he is not likely to pull it back. . . . There can be no thought of finishing,' von Braun had said. 'We have set sail upon . . . the cosmic sea of the universe. There can be no turning back,' agreed a later head of NASA.[10] Space travel, like progress itself, had only one direction: forwards.

This was not traditional political lobbying but a long-range project of cultural engineering designed to convince the public that mankind's future lay in space, and that that future was already arriving. Only then would taxpayers and Congress be likely to come up with the federal funding needed for a full-blown space programme. In 1949, only 15 per cent of Americans expected a Moon landing within fifty years. By 1960, a year before President Kennedy's famous pledge to go to the Moon within the decade, over half of the American public already expected a Moon landing within ten years.[11] Space travel had become real even before it happened.

Astrofuturism was given a specifically American spin by the 'frontier myth' – the idea that the core values and achievements of American society had been forged by the constant challenge of westward expansion.[12] The drive to find *Lebensraum* – living space – in the east had been the oppressive German version of the same ideal. But as the Second World War had demonstrated, the world was now full. Where could the restless spirits of Europe and America find room to exercise? With the assistance of Wernher von Braun the answer became obvious: *Lebensraum* in space.

In 1946 Arthur C. Clarke wrote: 'interplanetary travel is now the only form of "conquest and empire" compatible with civilization.'

Later, as the manned space age got underway, he explained that 'the opening of the space frontier' would cure the 'malaise [that] has gripped the western world . . . by providing an outlet for dangerously stifled energies'. President Ronald Reagan later put it like this: 'space, like freedom, is a limitless, never-ending frontier on which our citizens can prove that they are indeed Americans'.[13] It was no coincidence that one of the biggest hit films of the early space age was the 1962 epic *How the West Was Won*, told in triple-screen cinerama.

How the West Was Won in turn supplied part of the inspiration for another widescreen epic, Arthur C. Clarke and Stanley Kubrick's *2001: A Space Odyssey*. This, the ultimate manifesto of techno-futurism, appeared in the spring of 1968, just before Apollo got underway. Its realism was achieved through close contact with the actual space programme, which in many ways it anticipated right down to the taciturn astronauts and the beautiful and dramatic view of Earth at the end. Set in the tangible future, its theme was humankind's first contact with extraterrestrial intelligence, and its message was that space travel marked a decisive step in human evolution.[14] The phrase 'technological evolution' was beginning to be heard in these years, as if evolution were a cultural phenomenon, an act of will; it lent a useful air of inevitability to developments that as yet were only hoped for.

As Apollo 8 headed for the Moon, the mood was already historic. 'It is as thrilling as if we were riding ourselves in the crow's nest on the foremast of the Santa Maria,' wrote an excited curator at Boston Museum of Science.[15] For the British astronomer royal, Sir Bernard Lovell, normally sceptical about the benefits of manned space flight, it was 'one of the historic moments in the development of the human race'. For a week the press, radio and TV throughout Europe and the Middle East carried special stories and bulletins. Afterwards, American newspaper editors voted Apollo 8 the news story of the year, ahead of all the riots, wars and assassinations.[16] Having spent two decades making space seem real to the American public, astro-futurists now revelled in that reality; it was a euphoric moment. The author of *2001*, Arthur C. Clarke, wrote: 'the world that existed before Christmas 1968 has passed away as irrevocably as the Earth-centred universe of the Middle Ages. The second Copernican revolution is upon us, and with it, perhaps, the second Renaissance. . . . Many of the children born on the day that Apollo 8 splashed down may live to become citizens of the United Planets.'[17]

A few hours after that splashdown, NASA's chief, Thomas O. Paine, compared the voyage of Apollo 8 to that of Columbus and set out a long-range programme of space stations and planetary exploration. 'We are here this morning at the onset of a program of space flight that will extend through many generations,' he proclaimed. 'Man has started his drive out into the universe. It is the beginning of a movement that will never stop. One hundred thousand miles out from Earth there is no room for a space race, no place for Russian–American competition. This is something for all mankind.'[18]

The American press celebrated Apollo 8 in similar terms. 'The boundless frontier has been opened. Man's horizon now reaches to infinity,' proclaimed the Washington *Evening Star*. 'It boggles the mind,' said the *Los Angeles Times*. 'Man, after thousands of years of life on this planet, has broken the chains that bind him to Earth.' *Time* magazine pondered: 'man is propelled from Earth to Moon by the same instincts that led him from cave to college: the lonely search for knowledge, the fascination for attacking the impregnable, the creative impulse.' The semi-official Soviet congratulation declared that the voyage of Apollo 8 'goes beyond the limits of a national achievement and marks a stage in the development of the universal culture of Earthmen'. Addressing both houses of Congress the following month, Borman and his colleagues spoke once again of the 'overwhelming emotion' of seeing the Earth from space, but their final message was that 'exploration really is the essence of the human spirit, and to pause, to falter, to turn our back on the quest for knowledge, is to perish'. Their mission, they said, had been 'a triumph of all mankind'.[19]

Behind the scenes, however, the rhetoric was already beginning to turn stale. Two months before Apollo 8, Paine drafted a speech for his colleague Willis Shapley to be delivered to the Office of Naval Research. He filled in with shorthand: 'OK – then blah blah blah about the challenge of space . . . end on upbeat note . . . . And so today the spacefaring nations are acquiring a new cosmic view, a new and better understanding of man's position in the universe, as the challenge of space inspires us to blah blah blah.' For those in the stream of it, astrofuturist rhetoric was becoming a kind of Orwellian 'quackspeak', a prefabricated phrase-making that rolled off the tongue without the need for thought. The comparison with Columbus did not quite work either. Columbus had set out to rediscover an old continent, Asia, and had ended up discovering a new one, America. Apollo 8 set out to discover a new world, the Moon, and ended up rediscovering the old one. The astronauts, remarked the *Houston Chronicle*, belonged to 'a geographical unit we will be hearing about more in the future – Earth'.[20]

Some familiar earthbound criticisms of the space programme were heard again: there were 'still as many worlds to conquer at home as there are in space'; 'man can leap over the Moon . . . but he can't find a way to live at peace with his neighbours'; 'why cannot the same kind of mobilization of resources be utilised to meet the nation's real problems here on Earth?' 'Everything Earth-bound that cannot be done, everything Earth-bound that has not been understood, is made to seem a far greater failure when it is the failure of people who can touch the Moon,' thought a columnist in the Philadelphia *Evening Bulletin*. A cartoon in the *St. Louis Globe-Democrat* showed humanity pulling at the coat-tails of a Moon-struck scientist, asking, 'could I interest you in some Earthly problems?'[21]

On the whole, however, the view of the distant Earth inspired optimism. The *Christian Science Monitor*, pleased by the fulfilment of Mary Baker Eddy's prophecy that one day men would view the universe from beyond the Earth, put it like this: 'no man, no nation, no race can fail to think more broadly as a result of men's having circled the Moon. With such an achievement in their eyes, fewer persons will be tempted to believe that Earth's problems, however stark, are beyond settlement . . . the space programme's greatest and healthiest impact is almost certain to be on events back here on Earth.' Everywhere, newspaper editors wrote about the brotherhood of man and the spiritual unity of mankind.[22]

The weekly magazines had the advantage of being able to print the newly released photographs of Earth. *Time* advertised its end-of-year number with a photograph of Earthrise and the single word, 'Dawn'. *Life*, in a New Year issue read (it was claimed) by one in four Americans, presented a sumptuous photo-essay on the mission, with the Earth filling a cover bearing the headline: 'The Incredible

Year '68'. Inside was a poster-sized double-page spread of the Earthrise photo, and lines from the poet James Dickey: 'Behold/ The blue planet steeped in its dream/ Of reality.'[23]

One commentary stood out above all others: the poet Archibald MacLeish's essay 'Riders on the Earth'. 'For the first time in all of time,' he wrote, 'men have seen the Earth: seen it not as continents or oceans from the little distance of a hundred miles or two or three, but seen it from the depths of space; seen it whole and round and beautiful and small.' This view, he prophesied, would remake mankind's image of itself. 'To see the Earth as it truly is, small and blue and beautiful in that eternal silence where it floats, is to see ourselves as riders on the Earth together, brothers on that bright loveliness in the eternal cold – brothers who know that they are truly brothers.'[24] MacLeish's words featured in the *New York Times* on Christmas Day, as the TV pictures of Earth and the Genesis broadcast still resonated in the mind. They were widely quoted again a few days later when the Earthrise photo appeared, and were extensively syndicated across the American and British press. It was the single most widely admired evocation of the spirit of Apollo 8.

A newer strand of thought rose with the Earth: reverence for the environment. 'No man ever before has looked at the world in one piece and told us about it,' said the *Sunday Denver Post*. 'Perhaps with the new understanding will come reverence for our planetary home and for the uniqueness of life.' 'We should cherish our home planet,' advised the *Christian Science Monitor*. 'Men must conserve the Earth's resources. They must protect their planetary environment from spreading pollution. They have no other sanctuary in the solar system. This, perhaps, is the most pertinent message for all of us that the astronauts bring back from the Moon.'[25] Looking back, it is possible to see that Earthrise marked the tipping point, the moment when the sense of the space age flipped from what it meant for space to what it meant for the Earth.

A few far-sighted thinkers noticed this rising Earth-awareness quite early on. C. P. Snow suggested that Apollo, 'as well as being the greatest exploration . . . was very near the final one', and prophesied that civilisation would be 'driven inward' by it. 'How drab and grey, unappealing and insignificant, this planet would be without the radiance of life,' wrote the microbiologist René Dubos. 'I think the greatest contribution of Apollo has been to convert all those abstract ideas, like Spaceship Earth and global ecology, into an awareness that there is something unique about Earth and therefore something unique about man.' The biophysicist John Platt wrote: 'the great picture of Earth taken from the Moon is one of the most powerful images in the minds of men today and may be worth the cost of the whole Apollo project. It is changing our relationship to the Earth and to each other. I see that as a great landmark in exploration – to get away from the Earth to see it whole.'[26]

Fifteen months later, the US was celebrating the first Earth Day. Just beforehand, a correspondent to the magazine *Science*, John Caffrey, wrote: 'I date my own reawakening of interest in man's environment to the Apollo 8 mission and to the first clear photographs of Earth from that mission . . . I suspect that the greatest lasting benefit of the Apollo missions may be, if my hunch is correct, this sudden rush of inspiration to try to save this fragile environment – the whole one – if we still can.' Almost exactly four years after Apollo 8, the last of the Apollo missions brought back a still more famous photograph, the 'Blue marble' shot of the full Earth. It was, wrote the ecologist Donald Worster, 'a stunning revelation . . . . Its thin film of life . . . was far thinner and far more vulnerable than anyone had ever imagined.' Suddenly the image of the Earth was everywhere; it seemed to some to mark 'a new phase of civilisation', the beginning of the 'age of ecology'. It has been called 'the most influential environmental photograph ever taken'.[27]

Since then, the phrase 'blue planet' has come to be bound up with the idea of caring for the Earth. It has been used as the name of a long-running children's series on American public TV with an environmental theme, of a stunning British nature documentary series on the life of the oceans, and of a NASA programme to map every square kilometre of the Earth's environment from space, to name but three. Yet until the mid-1960s, no one really knew what colour the Earth would be. Imaginative pictures of the Earth from space, such as Chesley Bonestell's 1952 space station, show something rather like the traditional blue and green geographical globe, the land (usually north America) prominent and clearly defined, the oceans greenish, and the clouds optimistically few. When the whole Earth was finally photographed clearly there was surprise at the dominance of dazzling blue ocean, the jacket of cloud and the relative invisibility of the land and of human settlement. The sight of Earth seemed humbling, a rebuke to the vanity of humankind – just as ancient philosophers had foreseen.

The idea that environmental awareness has in some way been bound up with the sight of the Earth from space has been often proposed but rarely investigated; few people, and still fewer historians, are interested in both the environment and the space programme. Yet, as Andrew Smith has observed, 'there seem to have been two sharply delineated space programmes running parallel within the programme – an official one about engineering and flying and beating the Soviets, and an unofficial, almost clandestine other about people and their place in the universe; about consciousness, God, mind, life.'[28] It is this unofficial space programme that interests us now. Two remarkable films have brought home to a later generation the sheer magic of the first space age: Al Reinert's *For All Mankind* (1989), and David Sington's *In the Shadow of the Moon* (2007). Both focus on the experiences of the astronauts and both linger on the view of Earth from space.

That view was presented afresh to public attention in Michael Light's exhibition *Full Moon*. Put together for the thirtieth anniversary of the Moon landings, it displayed magnificently restored copies of some of the Apollo photographs, including room-sized panoramas of the lunar landscape. What prompted most comment among reviewers was that even in colour, the Moon was still black and white. One photograph was not enlarged. In a doorway between two parts of the exhibition was a small blob of colour floating over a bone-dead landscape. The caption was if anything more arresting than the picture: 'Earthrise, seen for the first time by human eyes.' 'Earthrise' alone was striking enough; 'seen for the first time' introduced a historical perspective; but why add 'by human eyes'? What other eyes might have seen this view, and how long ago? The perspective expanded again, to embrace all life in the universe, and all time since the Creation. The questions it raised lie behind this book.[29]

The first space age of 1957–72, from Sputnik to Skylab, is now very much part of the past, just like the super-decade of the 'long 1960s' that so neatly contains it. In the generation that followed, space history (with notable exceptions) tended to be an extension of that space age, practised by participants and specialist observers and restating and amplifying its goals and assumptions. Around the turn of the century, however, space history came of age as works began to appear which treated the ideals and assumptions of the first space age as part of the historical package, not necessarily rejecting them but seeing them critically and in context.[30] To gain a full perspective view of our subject we need to get outside it and view it from a distance, just as the Apollo astronauts viewed the Earth.

Earthrise is a history of the first views of Earth from space: how they came to be taken, what impact they had at the time and what their wider significance has turned out to be. The history of space and the history of the Earth need to be put back together where they belong. I have sought to combine interests in both space and the environment with the trade of the historian to produce a rounded account, based as far as possible on the record of what was seen, said and done at the time. The Apollo programme has been massively documented, not least through the excellent NASA Oral History Project, and I am not sure that a new round of interviews would bring us any closer to the period. I have consulted on a few important points with those still living, and have inevitably relied on memoirs written later on, but the focus remains on the first space age itself.[31]

The pursuit of the first pictures of the whole Earth has involved a long and varied journey taking in some unexpected places, so a brief summary of the book may be helpful. I have chosen to open it with the voyage of Apollo 8 at Christmas 1968, which is related afresh in chapter 2. It marked the apex of the space age – no subsequent manned mission has been further from Earth – and it gave us the revelation of Earthrise. The attention given to the first Moon landing of Apollo 11 a few months later has rather drowned out Apollo 8, and the story of this astonishing mission is well worth retelling. It is the story of how the mightiest shot in the Cold War turned into the twentieth century's ultimate utopian moment.[32]

Chapter 3, 'A short history of the whole Earth', stands in 1968 but takes a long look back at how people had imagined what the Earth would look like from afar, from the science fiction writers of the twentieth century right back to the philosophers of ancient times.

There are some prophetic moments; these changing (and sometimes familiar) expectations are a story in their own right. Apollo 8's view of the whole Earth seems to have had most in common with the most ancient visions of all.

If humans first saw the whole Earth in the space age, when did they first see that the Earth was round? Chapter 4, 'From landscape to planet', goes back to the postwar years – and, in fact, earlier still – to search for the first photographs to show the curvature of the Earth. It turns out too that the Apollo 8 mission was not the first time that the whole Earth was photographed. Chapter 5, 'Blue marble', goes in search of the earliest pictures of the whole Earth, taken from fragile lunar probes, tiny communications satellites, monstrous rockets and then, of course, by the men who went to the Moon. All these photographs were by-products of some different project, often the result of ingenious efforts, for throughout the 1960s very few people in NASA thought that pictures of the Earth were worth serious effort or expense. That we have them at all is largely down to the behind-the-scenes efforts of one man: Richard Underwood, who trained all of NASA's astronauts in photography and whose remarkable story deserves to be better known.

In the second half of the book we turn from the view of Earth to the impact it had: on those who beheld it, on those who saw the photographs and on our understanding of the Earth itself. Chapter 6, 'An astronaut's view of Earth', looks at the experiences of the people – only twenty-four in number – who actually left the Earth's orbit and saw it from afar. In the 1970s curious stories began to appear of astronauts who had become evangelical Christians, joined the 'New Age' movement, suffered mental breakdowns, gone into politics, or simply gone home to think. What comes over from their testimonies is that what produced such powerful effects was not just the experience of going to the Moon but going to the Moon and looking back at the Earth.

Just as Apollo 8 pulled back to view the Earth in the context of space, so we pull back to view Apollo 8 in the context of its time. Chapter 7, 'From Cold War to open skies', asks why it was that Apollo 8 and its view of the Earth made such a deep impact. In three areas the space programme seemed to move onto a new level in 1968: the awareness of Earth, the freedom of space, and the sense of the heavens. Chapter 8, 'From Spaceship Earth to Mother Earth', focuses on the biggest cultural change of all associated with the view of the whole Earth: the rise of environmentalism. An 'eco-renaissance' took place during the Apollo years of 1968–72, framed almost exactly by the 'Earthrise' and 'Blue marble' photographs four years apart. These were the years of the legendary *Whole Earth Catalog*, Friends of the Earth, and Earth Day, with environmentalism flowing into the anti-nuclear weapons crusade of the 1980s which also took the whole Earth as its emblem. These movements turned against the space programme, but they owed much of their early inspiration to its most important product: the image of the Earth.

Chapter 9, 'Gaia', looks at James Lovelock's Gaia hypothesis, that the Earth's living systems act together as a kind of super-organism that shapes its own planetary environment. The idea that the Earth is alive owed both its inspiration and its influence to the sight of Earth from space. This kind of holistic thinking about the Earth took root during the first space age and has been spreading gradually ever since, transforming our understanding of humankind's place in the universe more thoroughly than did the old astro-futurism. The end of the Cold War was accompanied by the rise of global warming. The years 1988–92 brought the Earth Summit, global citizenship, NASA's 'Mission to Earth', and new and more distant views of the home planet Earth as a 'Pale blue dot'. The final chapter gathers together the story of a quiet revolution that took place in the late twentieth century: the discovery of the Earth.

This book is about that extraordinary moment in 1968 when humankind first saw the whole Earth, and about everything that flowed into and out of it. It is an alternative history of the space age, written from a viewpoint looking back at the Earth. Confidence in the progress of science and technology was never higher than at the time of the first journeys to the Moon; afterwards came the first 'Earth Day', the crisis of confidence and the environmentalist renaissance. At the very apex of human progress the question was asked, 'Where next?', and the answer came, 'Home'. Earthrise was an epiphany in space.

CHAPTER 2

Apollo 8: from the Moon to the Earth

The short film clip of Neil Armstrong's 'one small step' onto the Moon during the Apollo 11 mission of July 1969 is often made to stand in for the whole of the space programme, just as the space programme itself has become a kind of shorthand for progress. Yet the first people to land on the Moon were not the first to visit it: that was the crew of Apollo 8, at Christmas-time 1968. It was, thought Michael Collins, 'magic . . . more awe-inspiring than landing on the Moon'. Collins was in a position to know: he was both a member of the Apollo 11 crew that performed the first Moon landing, and the main ground communicator ('capcom') with Apollo 8 from Mission Control, Houston. Of the two missions, he judged Apollo 8 the more significant.[1] And whereas the Apollo 11 mission had been thoroughly rehearsed in every detail except the actual landing, the Apollo 8 voyage was a launch into the unknown that provided a series of spectacular surprises. The greatest of these was the sight of Earth rising over the Moon. The impact of this serene vision was all the greater because of the risks and uncertainties of that most extraordinary of all space voyages.

Racing to the Moon

Until the mid-1960s, the Soviet Union had seemed to be ahead in the space race. It had achieved a string of firsts: the first satellite (Sputnik, in 1957), the first man in space (Yuri Gagarin, 1961), the first dual mission (1962), the first woman in space (Valentina Tereshkova, 1963), the first three-man mission (1964), the first spacewalk (1965), the first lunar soft landing (1965) and the first probe to land on another planet (Venus, 1966). As early as 1959 one Soviet probe had actually hit the Moon and another had photographed the far side. In the age of Khrushchev, achievements like these had fostered a widespread impression in the western world that the Soviet Union was rapidly catching up with the West, if not actually overtaking it in some areas, and that the future probably lay with some kind of progressive accommodation between socialism and capitalism. A lot was riding on the Apollo programme.

From 1965 America began visibly drawing ahead in space as the Gemini programme brought back stunningly clear photographs of orbital manoeuvres, space walks and the beautiful blue planet below. The Russians meanwhile ran into trouble, although the full extent of Soviet difficulties was a well-kept secret; the greatest of these was the death of the master rocket engineer Sergei Korolev, the Soviet equivalent of Wernher von Braun. For a gap of three years, from the spring of 1965 to the autumn of 1968, there was only one manned Soviet mission, and in that the cosmonaut was killed returning to Earth. The American manned programme had its own disaster when in January 1967 three experienced astronauts died in training on the launchpad of Apollo 1 as a flash fire ignited the oxygen in their capsule. Once the inquiries were over, however, NASA, desperate to meet Kennedy's end-of-the-decade target for a Moon landing, embarked on a period of 'high-risk, high-gain leadership'.[2]

In 1968 everything in America seemed to move faster. The assassination of Martin Luther King in April was followed by a destructive wave of urban rioting, defiant anti-Vietnam War protests, surges of student protest and, in June, the second Kennedy assassination, that of the presidential hopeful Robert Kennedy. As social programmes clamoured for funds and NASA fought off criticism and budget cuts, the Apollo programme went ahead at a pace that still seems scarcely believable.

Manned flight resumed at last in October 1968 when Apollo 7 took three men up to orbit the Earth for eleven days, thoroughly testing the craft that was to go to the Moon. The astronauts took a TV camera and broadcast regularly, establishing a daily public presence that was to be the mark of the Apollo flights. The Russians resumed manned missions a few days later with their first Soyuz flight; now both sides were back in business. Two weeks after that, on 12 November 1968, it was announced that Apollo 8 would not after all be another round of orbital tests: it was going to the Moon. The haste with which Apollo 8 was sent up had an important effect on the way the mission was set up; to it, we may owe the first Earthrise photograph.

Until August 1968 the plan had been that an 'A' mission would test the basic Apollo spacecraft (that was Apollo 7), then a 'B' mission would test the lunar module in Earth orbit, and finally a 'C' mission would test it in lunar orbit. After one further full dress rehearsal (the future Apollo 10), Apollo 11 would land on the Moon. The problem was that the lunar landing craft was not yet ready; there had been problems uncovered in the testing which would not be ironed out until well into 1969. If this A-B-C programme was adhered to, President Kennedy's all-important target of putting a man on the Moon before the end of the decade would probably be missed, turning Apollo from a success into a failure. In early August there was a flurry of activity at the top levels of NASA over a dramatic suggestion: bring the 'C' mission forward and go to the Moon soon, but without the lunar module.

This tied in with NASA's developing strategy of 'all-up testing', a concept so bold that at first even those in the know outside NASA did not take it seriously.[3] NASA's reasoning was that testing each system in turn would take time and, because of the way everything worked together, would not necessarily yield realistic results. The idea was now to send it all up at once to see how the system worked as a whole – all five million parts, with the astronauts riding on top. The first Apollo Moon mission would also be the first manned all-up test of the giant Saturn V rocket.

This was a risk: the Saturn V had only been tested twice before, unmanned. In the first test, designated Apollo 4, it had been sent looping 11,000 miles out from Earth. That test had been designated 'perfect'; it had also brought back wonderful colour photographs of the crescent Earth (as recounted in chapter 5 below). In the next test however, Apollo 6, the Saturn V had developed what the Apollo programme director General Sam Phillips admitted were 'several important technical failures and malfunctions'.[4] Three of the five engines had gone wrong and the rocket had started 'pogoing', producing vibrations that threatened its stability. According to Deke Slayton, head of the astronaut corps, 'it actually pointed itself down toward the centre of the Earth', then overcorrected, 'went into orbit thrusting backward, and wound up in an elliptical orbit rather than the circular one we needed'. Wernher von Braun's team of rocket engineers at the Marshall Space Flight Center in Huntsville, Alabama, set about investigating the mysterious pogo effect. They became confident that they had found the cause and that the vibrations could be cured by damping: Saturn V would be safe to use.

The crucial moment came when a deputation of all the senior engineers and managers on the Apollo programme, led by the director Sam Phillips, visited the Washington DC office of NASA's deputy head, Thomas Paine, to press their view that Apollo 8 should go all the way to the Moon. Paine reminded Phillips that up to that time there hadn't even been a manned Apollo mission: 'now you want to up the ante. Do you really want to do this, Sam?' 'Yes,' said Phillips. Every individual present backed him up, each laying his professional reputation on the line. Paine was impressed. Phillips then phoned right to the top. George Mueller, the head of manned space flight, the bold originator of 'all-up testing', was 'sceptical and cool'. NASA's chief, James Webb, was 'clearly shaken by the abrupt proposal and by the consequences of possible failure', but came cautiously to accept the consensus. Apollo 8 remained officially an Earth orbital test, but Phillips was authorised to mention a lunar option at the press conference; so deftly did he slip it in that hardly anyone noticed at the time.[5]

The first the mission commander, Frank Borman, heard about the change of plan was when he received an urgent call one Sunday from Deke Slayton: 'I can't talk over the phone. Grab an airplane on the double.' Borman flew a training plane to Houston and walked in to Slayton's office. 'I knew something was up when he asked me to close the door. "We just got word from the CIA that the Russians are planning a lunar fly-by before the end of the year. We want to change Apollo 8 from an Earth orbital to a lunar orbital flight. I know that doesn't give us much time, so I have to ask you: do you want to do it or not?" "Yes," I said promptly.' Borman had just volunteered his crew to fly to the Moon. He later explained: 'The Apollo program to me, was . . . a triumph in the battle of the cold war, that's why I was there. If you think that I would've devoted that much of my life simply to exploration or science, I wouldn't have, I'm not built that way, that's not my thing. But I was in the military, America means a lot to me, I believe in freedom, and from my standpoint, we were preserving it.'[6] Perhaps Slayton was also making a shrewd appeal to Borman's patriotism; if Webb did have secret intelligence about Russian intentions, wrote Slayton, 'damned little of it was shared with people at my level'.[7]

The decision to bring Apollo 8 forward and send it to the Moon has become one of the most minutely debated decisions in NASA's history, but it was undoubtedly driven by both technological and cold war factors. None of the records of the detailed discussions among NASA managers and engineers surrounding the decision show direct evidence that Russian intentions were a factor, but the overall urgency about the Apollo programme was dictated by Kennedy's end-of-the-decade target, which was very much about beating the Russians. NASA also had some recent information about the competition. Earlier in 1968 a mock-up of the giant N-1 rocket had been photographed on the launchpad at Baikonur, and a secret CIA report had claimed that the Soviet Union planned to send cosmonauts around the Moon later that year. At the time of the Apollo 8 decision, General Sam Phillips, the head of the Apollo programme, decided not to make a much-needed visit to meet James Webb, then at an international conference in Vienna, in order to avoid revealing to the Russians that something was up.[8] Webb's successor as NASA chief in September, Thomas Paine, slipped in 'a few comments on Soviet activities' as he briefed the outgoing President Johnson on the decision to send Apollo 8 to the Moon. Thereafter he ensured that he himself was fed the CIA's best evidence on Russian intentions in order 'to avoid major technological surprises'.[9] In September 1968 and again in November the Soviets sent a Zond spacecraft, their equivalent of Apollo, on unmanned voyages around the Moon, returning accurately to Earth with remarkable photographs of the Earth which in some respects rivalled those of Apollo 8.[10]

While the mission planners were told to keep the change of plan a secret, flight controller Gene Krantz recalled that 'it was like trying to hide an elephant in your bathtub'. On 12 November, NASA announced that the Apollo 7 mission had been 'a hundred-and-one per cent successful', and that it was time to put an Apollo capsule on top of a Saturn V rocket and send it out of Earth orbit entirely – leaving the unfinished lunar module back on Earth.[11] The first three men to travel to the Moon would be Frank Borman, the commander, focused and decisive; his jovial colleague from Gemini days James Lovell, the future commander of the ill-fated Apollo 13; and first-time astronaut Bill Anders, now a lunar module pilot without a lunar module. The launch date was only six weeks away: 21 December.

Behind the confident facade, the risks were known. Bill Anders reckoned there was 'one chance in three of a successful mission, one chance in three of an unsuccessful mission yet surviving, and one chance in three of an unsuccessful mission and not surviving . . . we couldn't get insurance'.[12] One Soviet cosmonaut wrote in his diary of how he and his colleagues were 'haunted' by the prospect of the American mission: 'we know everything about the Earth–Moon route, but we still don't think it is possible to send people on that route.' As Gene Krantz explained, the flight controllers were 'threading the needle, shooting a spacecraft from a rotating Earth at the leading edge of the Moon, a moving target a quarter of a million miles away'. The Saturn V rocket and Apollo spacecraft together had 5.6 million parts; even a 99.9 per cent reliability rate would leave 5,600 faults, any one of which might be fatal – as Apollo 13 was to find out.[13]

Still, the decision to send Apollo 8 to the Moon early meant that there would be no lunar module stuck to the front of the craft to block the view of Earth. It also meant that the astronauts would be a little less busy in lunar orbit and would have more time to look out of the window.

The flight of Apollo 8

The launch on 21 December was awesome beyond all experience. The writer Anne Morrow Lindbergh, among the crowds watching from three miles away, saw the rocket, burning 17 tons of fuel a second, rise 'slowly, as in a dream, so slowly it seemed to hang suspended on the cloud of fire and smoke'. The noise only arrived later, 'a shattering roar of explosions, a trip-hammer over one's head, under one's feet, through one's body. The Earth shakes, cars rattle, vibrations beat in the chest. A roll of thunder prolonged, prolonged, prolonged.' The aerial shockwaves were the strongest since Krakatoa. For Susan Borman, it was 'awesome . . . like watching the Empire State Building taking off'. Norman Mailer eventually found the words: 'now man had something with which to speak to God'.[14] Eleven minutes later Apollo 8 was quietly orbiting the Earth.

After three orbits the crew prepared to leave Earth altogether. Looking back, Michael Collins, the capsule communicator (or 'capcom'), felt that NASA hadn't really risen to the occasion.

> A guy with a radio transmitter in his hand is going to tell the first three human beings they can leave the gravitational field of Earth, what is he going to say? He's going to invoke Christopher Columbus or a primordial reptile coming up out of the swamps onto dry land for the first time, or he's going to go back through the sweep of history and say something very, very meaningful . . . I can remember at the time thinking, 'Jeez, there's got to be a better way of saying this,' but we had our technical jargon, and so I said, 'Apollo 8, you're go for TLI.'

'TLI' stood for 'trans-lunar insertion', the third stage rocket burn that would fire them out of Earth orbit. 'A hush fell over Mission Control . . . . After TLI there would be three men in the solar system who would have to be . . . considered a separate planet.' 'When I heard the crew report the manoeuver's completion it really hit me,' recalled Gene Krantz. 'I had to get up and walk outside because I was so happy I was crying.'[15]

'When we looked back,' remembered James Lovell, 'we could actually see the Earth start to shrink.' They could see the Earth because the craft travelled booster end forward in order to fire the engine to brake it into lunar orbit, leaving the cabin windows facing the Earth. 'I don't want to see you guys looking out the window,' Borman had warned, but of course 'everybody wanted to look out of the windows'.

> All of us had flown airplanes many times and seen airfields and buildings getting smaller as we climbed. But now it was the whole globe receding in size, dwindling until it became a disk. We were the first humans to see the world in its majestic totality, an intensely emotional experience for each of us. We said nothing to each other, but I was sure our thoughts were identical – of our families on that spinning globe. And maybe we shared another thought I had . . . . *This must be what God sees.*[16]

The onboard computer, tiny by comparison with modern pocket calculators, radioed their position back to Houston; the answer came back that their course was accurate to within two thousandths of a degree. The only course corrections required for the rest of the outward journey amounted to barely more than a delicate nudge to counteract the effect of jettisoning urine out of the side of the ship. After all the fire and thunder, Apollo 8 simply floated to the Moon. In a quiet passage, Collins relayed a question from his five-year-old son: 'Who's driving?' 'Isaac Newton is driving,' explained Lovell. Having been launched with all the urgency of the space race, Apollo 8 now seemed to have all the time in the world.

Digital photography still lay well in the future, so the first pictures of Earth to reach home were the TV transmissions, in black and white. Borman had not wanted to take a TV camera at all, seeing it as a risky distraction; the crew of Apollo 7 had got fed up with staging 'dog-and-pony shows' and turned theirs off.[17] Mankind's first live view of the Earth came on Sunday, 22 December from 120,000 miles up – or, rather, out. After a wave or two from the crew the camera switched to Earth. 'It's a beautiful, beautiful view,' explained Borman, 'a predominantly blue background and just huge covers of white cloud.' There followed several frustrating minutes of blank screen as they struggled in vain to fit the telephoto lens. In the end they abandoned the attempt and filmed out of the hazy porthole using the same low light lens used for filming inside the craft. Viewers were assured that the fuzzy blob of light they were seeing was indeed the Earth, and the transmission soon reverted to floating toothbrushes and the like.

The crew's second television transmission, on Monday 23 December, provided the first recognisable view of the whole Earth. It was timed to mark the point when the craft moved from the gravitational pull of the distant Earth into that of the approaching Moon. By this time the telephoto lens had been fixed and the camera was ready pointing at the Earth. Again things began with a blank screen as Anders struggled to frame the Earth inside (as he put it) 'a 9-degree field-of-view lens with no way of aiming it except for looking down the side or putting some chewing gum on the top'. A corner of the Earth's disc drifted into view and then out again, and much of the commentary consisted of 'up a bit, down a bit' type instructions from Houston, made hard to follow by the 1.3 second time delay in transmissions: 'we always seemed to lag just that little bit behind instructions from the ground to move the camera this way or that,' explained Lovell.[18] Finally the camera settled on the Earth, two-thirds illuminated and lying serenely on its back. 'There she is, floating in space!' exclaimed the newscaster Walter Cronkite. 'You are looking at yourselves from 180,000 miles out in space,' announced Anders.

The Americas were in view and the astronauts tried to describe the outline of the continents below the cloud, but they were seeing the picture a different way round from those on Earth and it was hard to follow. Houston asked whether the crew themselves could see anything on the dark part of the Earth; negative, came the reply, 'this Earth is just too bright'. 'What I keep imagining is if I am some lonely traveller from another planet,' pondered James Lovell out loud, 'what I would think about this Earth, at this altitude – whether I think it would be inhabited or not.' 'Don't see anybody waving, is that what you are saying?' replied Houston laconically. 'I had seen previous photos of the whole Earth', recalled Lovell much later. 'The photography was poor and I was not impressed. An Earth globe gave me a better idea of what I would see from the Moon. It is when you see the Earth first hand from space that the concept of the Earth and its place in the solar system is apparent.'[19] The entire human race was in the frame, bar the three behind the camera. Looking down upon the Earth, the astronauts had achieved the dream of philosophers over the centuries; it seemed scarcely believable.

Nearly three days into the mission, Apollo 8 prepared to swing round the back of the Moon. With no Sun, and no glare from the Earth, suddenly the stars came out. Anders recalled: 'As I turned around and looked over my shoulders . . . there was a line, a curved line of no stars, of an immense black hole, and the hair sort of stood up on my neck, and I realized . . . that was the dark hole of the Moon; and I had a kind of a feeling of falling into a dark hole.' They had arrived facing back towards Earth, and this was the first time they had actually seen the Moon. The crew had to master themselves and prepare for the vital engine burn; one mistake and they would be lost in space without hope of recovery. Then, recalled Anders, came another distraction. 'Just as we were in the middle of this countdown with not very many seconds to go, something caught my eye out of my window . . . early lunar sunrise . . . we were seeing light on the Moon for the first time.' As commander, Borman brusquely directed Anders' attention to the job in hand. The engine burn was perfect. Interviewed on BBC television not long afterwards, Borman recounted: 'we went all the way to the Moon without ever seeing it. We made the first burn without seeing it and then we looked down and 60 miles away, there was the Moon. Right at the moment that they predicted we would lose radio contact, at the exact second, we did.'[20] Out of touch with the Earth, they were now more alone than any human beings in history.

For the next day or so Apollo 8 orbited the Moon every two hours, losing radio contact every time they went round the back. The flight plan, finalised a month before, specified that they spend their time on 'landmark familiarisation', 'landing site sighting' and 'dark side photography' – that is, looking at the Moon. Anders, the lunar module pilot with no lunar module, was the designated mission photographer. He had been briefed about what to look for geologically: 'the photography was so canned that they had the f-stops calculated from the lunar albedo, based on our longitude around the Moon, which I could basically do with my watch. I had it worked out for every revolution.'[21] It was now Christmas Eve, and around breakfast time on the east coast of America, when, on their second orbit, they made their first broadcast from the Moon, sending back eerie close-up television pictures of its surface. The Moon, said Lovell, was 'essentially grey, no colour . . . like plaster of Paris or dirty beach sand'. 'We had been trained to look at the Moon, we hadn't been trained to look back at the Earth,' recalled Anders.

The Earth wasn't easy to see. The capsule had only five small windows, three of which had become fogged by gas leaking from the seals, and it was still pointing backwards, away from the rising Earth. After three orbits the spacecraft flipped round as it settled into a lower orbit; now at last they were facing forward as they passed behind the Moon. As they came out from behind the Moon for the fourth time, they saw the Earth rise – the scene recounted at the start of this book.

The recording of the crew's words as they saw the Earth rise suggests that it came as a complete surprise, and this is the view taken by some of the standard histories of the mission. But when the then Apollo director of photography Richard Underwood came across the claim in a book review, he wrote to NASA to set the record straight:

Absolutely and totally false. Hours were spent with the lunar crews, including [the] Apollo 8 crew, in briefing on exactly how to set up the camera, which film to use, F-stop, shutter speed, etc., times of Earthrise and Earthset, where it would rise above lunar horizon, etc. These briefings were most comprehensive. I even made trips to Cape Canaveral two days before the flight to give a final briefing on it. We also wanted a series of photos as they moved away from Earth at specific times, for continuity record photos en route to the Moon.

The official photographic operations plan, however, was not promising. Among the low priority 'targets of opportunity' was to 'photograph the Moon and Earth from various translunar/transearth distances', for purposes including 'weather and terrain analysis with global coverage and from a long distance perspective', 'horizon and high atmosphere studies', 'Earth terminator studies' and, finally, 'general interest'.[22]

At the press conference held on 12 November to announce the mission, all eyes were, understandably, on the Moon; no one mentioned seeing the Earth. Behind the scenes, however, recalled Underwood, 'there were some battles going on within NASA, especially during lunar orbit. I argued hard for a shot of Earthrise, and we had impressed upon the astronauts that we definitely wanted it.'[23] The crew certainly understood the historic importance of the mission. At the pre-launch press conference they were asked by a NASA public affairs officer, 'What are your plans [for] trying to look out of the window?' 'I hope we can get some good views of the Earth from the Moon, and of the Moon trans-Earth,' Borman replied. 'Other than that, I'm not certain.' The crew were asked what part of the mission they were most looking forward to. 'The fact that we can see Earth set and Earth rise,' said Lovell.

All the same, when that moment came, as the recording shows and as Frank Borman has confirmed, they were 'taken by surprise', having been 'too busy with lunar observations' during the first three orbits. Borman saw the Earth rise first and took a black-and-white photograph with the only camera readily to hand, the one Anders had been using to photograph the Moon. Anders joked that the picture was not scheduled, well aware that his commander had until then been determined to keep the crew firmly focused on the flight plan and away from the windows. Both men asked Lovell for colour film, while Lovell himself, the most experienced photographer among them, became so excited he had to be told to calm down; 'we all wanted to take a picture,' he remembered.[24] It was Bill Anders who took the famous colour 'Earthrise'; happily, his window was clear. As they swooped back into radio contact with Earth they said nothing about what they had seen, but the taped record of the Earthrise moment from the on-board voice recorder was routinely uploaded to Mission Control. The experience was still fresh in their minds when they came to make the Genesis broadcast later that day, allowing the public to share in their 'God's-eye view' of the cosmos.

The crew left talking about the Earth until their second Christmas Eve broadcast, at 9:30 p.m. This had been decided in advance on advice from the US Information Agency, but as it turned out it also gave them time to absorb the experience of seeing the Earth rise.[25] This time the broadcast opened with a view of a glowing white dot above the pale lunar surface: 'a view of the Earth as we've been looking at it for the last sixteen hours,' explained Borman. The rest of the broadcast was devoted mainly to live pictures of the lunar surface as it drifted by the cabin window. 'A vast lonely expanse of nothing' was how Borman described it. He then invited his colleagues to give their personal reactions. 'A vastness of black and white, absolutely no color,' agreed Lovell. 'The loneliness out here is awe-inspiring. It makes us realize what you have back on Earth. The Earth is a grand oasis in the vastness of space.'

With only two minutes left before radio contact was lost, the moment had arrived for their prepared statement. Anders' voice altered in tone. 'We are now approaching lunar sunrise, and for all the people back on Earth, the crew of Apollo 8 have a message that we would like to send to you.' There was a pause: almost no one on Earth knew what was coming next. To flight controller Gene Krantz, sitting at his console in Houston, 'it was a surprise, beautiful and timely . . . I felt a chill as Anders said, softly':

In the beginning, God created the heaven and the Earth; and the Earth was without form and void, and darkness was upon the face of the deep; and the spirit of God moved upon the face of the waters. And God said, 'Let there be light,' and there was light.

The TV camera showed the shadows below lengthening as Apollo 8 flew into the lunar sunrise. Lovell took over to read the next section, then Borman signed off with fine assurance:

'And God called the dry land Earth, and the gathering together of the waters called He seas. And God saw that it was good.' And from the crew of Apollo 8, we close with good night, good luck, merry Christmas, and God bless all of you – all of you on the good Earth.

Anders promptly turned off the transmitter; seconds later Apollo 8 dipped once more behind the Moon, leaving the Earth in radio silence to absorb the impact of the words. 'For those moments I felt the presence of creation and the Creator,' said Krantz. 'Tears were on my cheeks.'[26]

The writer William Styron was in Connecticut, 'in a house filled with noisy festivity'. His host, a teacher, a hardened space sceptic, had to be persuaded to switch on the TV for the live pictures of the Moon. Then, recalled Styron, 'the murmur and laughter of the party diminished and died, and we watched in silence'. At the reading of Genesis, 'a chill coursed down my back and an odd sigh went through the gathering like a tremor . . . I glanced at my host, the mistrusting and scornful teacher, and saw on his face an emotion that was depthless and inexpressible.'[27] Such scenes were multiplied across the United States on Christmas Eve, 1968.

All that was left for the astronauts to do before they left the Moon was to perform the all-important engine burn to send them out of lunar orbit and towards home – exactly two minutes and eighteen seconds. It had to be done manually while out of contact with Earth; if they got it wrong they would never get back. As Christmas Eve passed into Christmas Day, the crew's family and colleagues waited anxiously for radio contact to resume. To the exact second, James Lovell's voice was heard. The burn too had been spot-on, so much so that a mid-journey course correction was abandoned; the accuracy was compared to shooting a letter through a letter-box from four miles away.

Like all long journeys home, Apollo 8's was an anti-climax. 'We fell for 240,000 miles really,' recalled Anders. 'Frankly, it was kind of boring.' But then Lovell (whose nickname was 'shaky', and who had earlier inflated his space suit by mistake) accidentally erased all the navigational information on the onboard computer, which promptly concluded they were back on the launch pad and shut up shop.[28] Since safe re-entry into the Earth's atmosphere depended on getting the angle exactly right, their fate hung upon Lovell's skill with a sextant as he made fresh star sightings. The figures were radioed to Earth for calculation, and the results read back by Mission Control to be fed digit by digit into the ship's computer. The astronauts' rigorous training in old-fashioned navigation and star recognition paid off; as the capsule approached the Earth, only the tiniest of adjustments was needed.

There had been nothing like Apollo 8's re-entry into the Earth's atmosphere before. Dipping down from Earth orbit was one thing, but plummeting in at 25,000 miles an hour from a quarter of a million miles up was quite another. They had to skip the craft in and out of the atmosphere to slow it down, like a pebble skimming on a lake. Of the four unmanned attempts (American and Russian) to bring unmanned craft down from beyond Earth orbit, two had failed. Too shallow an angle and the craft would bounce off the atmosphere into space, beyond all hope of rescue; too steep and it would burn up like a meteor. The distance between life and death was half a degree. The forces on the astronauts rose to over 6G and the temperature outside to 2,800 degrees Celsius, bathing the interior of the capsule in a neon glow. While all this took place the radio blacked out. Blackbird supersonic planes, skimming the upper atmosphere, could see the fireball but had no way of knowing what, if anything, survived inside it. Two Pan American airline pilots, flying in and out of Honolulu, saw a red ball of fire followed by a streamer a hundred miles long, like 'an orange slash in a piece of black velvet'.[29] From ships and helicopters waiting in the drop zone, binoculars were trained on the night sky, while the mission controllers waited to see if the crew were still alive.

After an agonising delay the radio fizzed back to life: 'we are in real good shape, Houston.' The parachutes deployed and a few minutes later Apollo 8 splashed down into the Pacific night. They passed right over the aircraft carrier waiting to pick them up, in some danger of actually hitting it, and landed less than three miles away. By now capcom Michael Collins was 'a basket case, emotionally wrung out'.

The emotional outpouring at Mission Control was intense: 'we grabbed for the lunar prize, and we got it on our first shot,' exulted Krantz.[30] When the divers from the USS *Yorktown* arrived at the capsule they took first not the astronauts but the film. As the mission controllers slept off the celebrations, NASA's photographic team set to work to see what the cameras had brought back.

Earthrise

The photographic processing for Apollo was done at NASA's Manned Spaceflight Center at Houston – home of Mission Control – by NASA's technical monitor of Apollo photography, Richard Underwood, who took great pains with the work.

> I prepared lists on everything . . . and I worked day and night to get them out so people around the world could have them. It was a labour of love . . . I wanted to get this information out to the world, so they could use these photographs to study various things. So the rolls came back and we gave them very tender loving processing in a very slow process.[31]

Automatic machinery could have done the job in minutes, but Underwood insisted on using a manual process which took five hours, so that if anything went wrong the film could be quickly rescued. The original negatives, virtually untouched, were used to make masters and then filed away; these masters, carefully numbered, were then used to the make prints. NASA's policy was (and is) to allow free use of its photographs: after all, taxpayers had paid for them, and NASA's federal budget depended upon taxpayers' appreciation and support. After processing, the prints were spread out along tables at the Manned Spaceflight Center and the returning crew gathered with NASA staff to help identify and select the photographs. A batch of briskly titled pictures was chosen for release through the Public Affairs Office.

'I had a pretty good idea of what it [Earthrise] was going to look like,' recalled Underwood, 'but when I actually saw the picture, after they returned, it was even better than I had anticipated.'[32] The captions supplied by the photographic service were, however, informative rather than poetic.

Apollo 8 Earth View. This view of the rising Earth was seen by the Apollo 8 prime crew, Astronauts Frank Borman, commander, James A. Lovell, Jr., command pilot, and William A. Anders, lunar module pilot, during their orbital flight around the Moon. The Earth is approximately five degrees above the lunar horizon. This photograph was taken while the spacecraft was 110° east longitude. The horizon, about 570 kilometers (350 statute miles) from the spacecraft, is near the eastern limb of the Moon as viewed from Earth. Width of view at the horizon is about 150 kilometres (95 statute miles). On Earth, the sunset terminator crosses Africa. The south pole is the white area near the left of the terminator. North and South America are under clouds. The lunar surface photography has less pronounced color than indicated in this photograph.[33]

A later version of the caption, amended by the Public Affairs Office, tried harder: 'This view of the rising Earth greeted the Apollo 8 astronauts as they came from behind the Moon after the lunar orbit insertion burn.'[34] The term 'Earthrise' had yet to be coined. But although no one noticed, there was something not quite right about the picture. Apollo 8 had been on an equatorial orbit with respect both to the Earth and the Moon – that is, it had been circling in the plane of the equator, as it were horizontally. It then went into orbit round the Moon in the same plane, going clockwise. As the astronauts saw it, Earth had not exactly 'risen' but had appeared around the left side of the Moon, North Pole at the top with the sunset line running vertically from north to south. In the original photo, too, the Earth had been a much smaller part of the dark sky; this was cropped in the version released to the public, making the Earth seem larger. Instinctively, the photograph had been altered from a Moon to an Earth perspective.

Any sense that something is wrong with a top-lit Earth is overwhelmed by the naturalness of seeing the Earth 'rise' over the Moon, with the lunar horizon below, in the same way the Moon rises above the Earth. The slight tilt of the lunar horizon adds drama without changing the basic orientation; it could easily be the camera that is tilted. In a sense, all orientations in space are merely conventions, for without gravity there is no tangible 'up' or 'down'. The official report prints the photos as they came, sideways, upside-down and at an angle. This may be partly explained by the lack of a viewfinder on the cameras, as Underwood explained: 'In the Mercury, Gemini, Apollo [capsules], you couldn't distort your body very well to look out the window of the spacecraft through a viewfinder, so they were taught to shoot from the hip.' Anders always mounted his own copy vertically, for this was how he had seen it at the time.[35]

There was one other way in which the photograph was not quite what it seemed to be: it was not actually the first. For a long time there was a dispute between Borman and Anders over who had taken the photo, with both men claiming it. Both were equally confident and equally trustworthy, which left a puzzle. It was eventually resolved by Robert Zimmerman, who realised that a black-and-white version had been taken first, showing the Earth low on the horizon. 'When Borman saw the Earth rising,' concluded Zimmerman, 'he immediately grabbed the camera Anders had been using and snapped a picture . . . Borman then handed the camera to Anders, who unloaded the black-and-white magazine, inserted the color one and began shooting.' So the very first Earthrise photo was Borman's black-and-white one, but it was overlooked for thirty years in favour of Anders' colour one taken shortly afterwards.[36] And, of course, while it was indeed 'Earthrise, seen for the first time by human eyes', there was no record of the three Earthrises that had gone by unnoticed.

Return to Earth

On 26 December 1968, even before Apollo 8 had returned, the acting head of NASA, Thomas O. Paine, wrote to President Johnson in euphoric terms. 'It is apparent that an unprecedented wave of popular enthusiasm for the Apollo 8 astronauts is building up around the world,' he said. 'Laudatory editorials are in every paper.' Even hardened press people and critics were talking of it as the greatest adventure in American, or even human, history. Messages of congratulation flooded in. When it was all over, Paine judged that the reaction had been 'unprecedented in the history of the space program', beyond even that accorded to John Glenn's orbital flight. 'Never have I seen such continuing public interest over an extended period nor a more complete job in communicating to the public,' commented Tom

Morrow of Chrysler.[37] It was estimated that over one billion people – nearly a quarter of the Earth's population – had seen some TV coverage of the flight, thanks to communications satellites which had only recently been put in place. Among them was NASA's ATS satellite, which had taken the first television pictures of the whole Earth, and Intelsat-3A, launched ahead of schedule to cover the mission. As the Washington *Sunday Star* put it, 'the impact was immediate, total, and worldwide. We were there.'[38]

It is hard to say which made the greater immediate impact: the reports of the first travellers to the Moon or the first pictures of the Earth. Before the mission, the Earth had not been in the frame, but it came from nowhere to command equal attention. On Christmas Eve, the US morning papers were able to report the crew's TV pictures of the distant Earth. Of twenty-three front pages collected in NASA's daily *Current News* digest, thirteen showed the Earth only, five the Moon only, and five both. On Christmas Day it was all Moon pictures as the papers reported on the broadcasts from lunar orbit, including the Genesis reading; there were no new pictures of the Earth, but much talk about it. On 28 December a number of papers carried pictures of the Earth again, filmed on the way home, and during the return journey many editors and columnists mused on the meaning of it all.

The Apollo 8 photographs of both Earth and Moon, including the famous Earthrise, appeared in the American press on 30 December. Despite their much better quality they marked the end of the wave of comment; most newspapers had already done their philosophising. Earthrise appeared on only six of NASA's collection of front pages compared with eighteen for the earlier TV picture. The *Washington Post* printed the Earthrise photo across the top half of its front page, headlined: 'From the Moon, Man Sees the Shining Earth'. The main public showing for Earthrise (as we saw in chapter 1) came in colour, in the weekly and monthly magazines. A little later, the *Houston Chronicle* printed it alongside another photograph taken by two of the same astronauts, Lovell and Borman, on Gemini 7, showing the Moon above the Earth's horizon. In Britain, the colour photographs of the Earth were shown on late night BBC news as soon as they came through, and excitedly reported on in *The Times* the next day: they were, thought the paper, 'a humbling reminder of the world's insignificance'. A few days later *The Times* went into colour to print four pages of photographs from Apollo 8, led by a full-page Earthrise.[39]

Like the photograph itself, the commentaries faced two ways at once: out into space and back at the Earth. There were some Earth-focused pessimists and rather more space-focused optimists, but the largest block of opinion seems to have been in the middle, enthralled by the sight of the home planet but at the same time excited by the idea that the forces which took men to the Moon could transform human life on Earth as well as in space. The *Los Angeles Times* summed it all up nicely:

> In retrospect, a remarkable effect of the Apollo 8 Moon voyage was not so much its capacity to draw men's gaze outward as its powerful force in turning their thoughts inward on their own condition and that of their troubled planet. . . . The flight of the astronauts produced great mental and spiritual ferment among men. The feat that should have been the perfect object for extro-verts made introverts of us all.

Religious people, observed the paper's bureau chief, no longer felt threatened by science; rather, the voyage was taken as 'a reaffirma-tion of the wonders God has wrought in the universe and of the divine spark in man'. The piece concluded with the hope that the United States would now be able to bring about 'the comity of people of different races and nationalities' and avert conflict in a nuclear armed world.[40]

When the Earthrise photo itself appeared the next day, it fitted the mood perfectly. Earth was seen in a new context in space, but at the same time appeared colourful and unique. This was where the appeal of MacLeish's 'Riders on the Earth' commentary lay: in its eloquent combination of forward-looking optimism and Earth-centredness. Subtly and surely, MacLeish's words, like Apollo 8 itself, caught the wonder of space and focused it on the Earth.

The formal celebrations for Apollo 8 opened on 9 January when the crew paraded through Washington DC, addressed both houses of Congress, and met the President. They presented him with a framed copy of Earthrise; later, they presented another copy to Governor

John Connally of Texas, a surviving passenger from the car in which John F. Kennedy had been assassinated. They invoked both the eternal spirit of exploration and the view of Earth from space, and concluded by quoting MacLeish's description of humankind as 'riders on the Earth together'. The next day there was a parade in New York and a state dinner with the mayor and the Roman Catholic archbishop. They were welcomed to the UN Security Council by its Secretary-General, U Thant, who described them as 'the first universalists'. 'We saw the Earth the size of a quarter, and we recognized then that there really is one world,' Borman responded. 'Apollo 8 was a triumph for all mankind.' At his inauguration as President the following week, Richard Nixon coupled the ritual vow to bring peace to the world with a quotation from MacLeish and a pledge to space exploration: 'as we explore the reaches of space, let us go to the new worlds together – not as new worlds to be conquered, but as a new adventure to be shared.' He staged a reception for the astronauts, where it was announced that (Lovell and Anders being back in training as back-up crew for Apollo 11) Frank Borman would shortly begin a goodwill tour of Europe.[41]

Borman's European tour in early February was a sensation. 'The overwhelming impression of the people of Europe was this view that we got of Earth,' he reported on his return home. 'They responded to the fact that we are really riders on the Earth together. It's small and beautiful and fragile.' In London he met the Queen and the Prime Minister, addressed the Royal Society and sat in the public gallery of the House of Commons to hear Apollo 8 lauded by MPs. In Paris he was presented with an original copy of Jules Verne's *From the Earth to the Moon* by the author's grandson, who pointed out the remarkable parallels between the two voyages a century apart. The visit to Brussels, home of European regulation, was marked by an announcement that the International Standards Organization was to meet to discuss harmonising the arrangements for space docking. He visited the Berlin Wall, which he described as 'tragic'; he had last been in the city as a serviceman during the 1948 airlift. In Bonn he spoke to an audience of space scientists which included Hermann Oberth, Wernher von Braun's former mentor, telling them, 'we are first and foremost not Germans or Russians or Americans but Earthmen'.[42]

In Rome, Borman showed a film of Apollo 8 to assembled cardinals and members of the Pontifical Academy of Sciences and then addressed them from the very spot on which Galileo had been condemned for the heresy of Copernicanism, saying: 'the most indelible image that remains in my mind's eye is the wonderful view of Earth. . . . National boundaries and artificial barriers that separate countries were invisible.' He then enjoyed a seventeen-minute audience with the Pope, unprecedented for someone who was not the head of a state or a Church. Borman returned a papal medal which had been around the Moon, and received from Pope Paul VI (an astronomy enthusiast) a number of souvenirs including copies of two ancient Bibles. The Pope praised the Genesis reading: 'For that particular moment of time,' he said softly, 'the world was at peace.'[43] Finally, Borman flew to Madrid and laid a wreath at the statue of Christopher Columbus. Everywhere he took the same 'one world' message about the need for peaceful coexistence and international brotherhood, a familiar sentiment given force by the evidence of his own eyes. 'If by leaving the Earth, we can help bring these [truths] forth,' he concluded, 'then the journey is one the world needs.'[44]

One visit had been put off, because of international tensions: Russia. During the flight of Apollo 8 Soviet TV had taken the unprecedented step of transmitting American TV coverage, and afterwards a telegram of congratulation arrived at NASA from the Soviet Academy of Sciences: 'Let the rockets work for peace uniting but not separating peoples.' During Borman's European tour Hungary had issued a commemorative stamp for Apollo 8, beating the American one by several months.[45] Borman finally made it to Russia in July, just before the Apollo 11 Moon landing, in a carefully managed tour which coincided with preliminary talks on arms control. Accompanied by Titov, the cosmonaut who had joked that he had not found God in space, Borman became the first American to visit 'Star City', the Soviet equivalent of NASA's Manned Spaceflight Center in Houston. He gave the cosmonauts a colour film of the voyage of Apollo 8, and his theme once again was that space exploration was useful 'to the entire planet,' not just individual nations. Borman's Russian visit, made possible by the universal significance of Apollo 8, was the first public sign of a phase of space cooperation that was to lead to the Apollo–Soyuz programme in the coming years of detente.[46]

Ever since it was released, the Earthrise picture has been associated with Archibald MacLeish's short essay 'Riders on the Earth'; the two have frequently been reproduced together. Although true for the period after the return of Apollo 8, however, when we approach closely the association turns out to be a historical trick of the light. While the astronauts had already seen the Earth rise, MacLeish had not. The piece first appeared on Christmas Day, while the undeveloped negatives of the Earthrise photographs were still orbiting the Moon. It must also have been written before the Genesis broadcast late on Christmas Eve, at a time when only the fuzzy black-and-white TV pictures of the Earth had reached home – and perhaps even before then. The astronaut Russell Schweickart, profoundly affected by his own experience of viewing the Earth, later wrote: 'you marvel that an Archibald MacLeish somehow knew that. How did he know that? That's a miracle.'[47] How indeed? For the answer to that question, and indeed to the question of why Apollo 8 made such a deep impact, we need to explore the historical context in which these remarkable events took place.

CHAPTER 3

A short history of the whole Earth

'What will we see when we leave the Earth?' The question was asked in *Universe*, a Canadian public information film made in 1960.[1] The film was designed to introduce the public gently to the discoveries of astronomers about the immense scale of the cosmos. In answer to the question, the viewer was shown an animation of the Earth appearing from behind the Moon, not above it but to the left, as Apollo 8 was to see it a few years later. This really was the view from space.

Universe was an important influence on the film director Stanley Kubrick, who adopted some of its effects a few years later in his own vision of the future, *2001: A Space Odyssey* (1968), the film he developed in the mid-1960s with Arthur C. Clarke. It too was intended to inform and prepare the public for discoveries to come, and it too anchored its narrative in a compelling view of the Earth from space. At the end of the novel *2001: A Space Odyssey*, Clarke described the Earth as seen by the 'Star Child', the newly evolved embryo of the next human race: 'There before him, a glittering toy no Star Child could resist, floated the planet Earth with all its peoples.' Here, the sight of Earth from afar marked the end of an era: 'history as men knew it would be drawing to a close.' The film itself closes with a prophetic scene. Drifting in space a human foetus, its eyes open, gazes with the blank wisdom of eternity upon the whole Earth. It is, we understand, an astronaut transformed by the experience of space travel. Nine months later the crew of Apollo 8 underwent a similar experience. They had attended a premier of *2001: A Space Odyssey* in Houston about three months before the flight; as they saw the Earth receding behind them, Anders thought about the film.[2]

In the astrofuturist framework, to look back on Earth from the outside was to look back on humanity's past from its future. Four centuries ago, Copernicus had proved that the Earth was not at the centre of the universe; this would finally sink in when the Earth appeared in the corner of the frame. When actual pictures appeared a few years later, however, they seemed to put the Earth back at the centre of attention, a view reinforced by the descriptions of the astronauts themselves. Earth certainly appeared small and insignificant from space, but (as Kubrick sensed in *2001*) this had the effect of focusing thoughts upon it, not away from it.

There was in fact a long historical tradition in which the imaginary vision of the Earth was an aid to thought – humbling thought – about the place of humankind in the universe. In a ground-breaking work, *Apollo's Eye*, 'a cultural history of imagining, seeing and representing the globe', Denis Cosgrove has shown how 'the meanings of the photographed Earth were anticipated long before the photographs themselves were taken'. 'Representations of the globe', Cosgrove argues, 'have exercised an especially powerful grasp on the western imagination.'[3] Just how powerful can be seen if we compare the insights of the techno-prophets and science fiction writers of the modern age to those of earlier times. Before the whole Earth was seen in reality, it appeared in imagination to a handful of far-sighted visionaries.

The view from the future

In the generation or so before the first space age, science fiction was at its peak of popularity. But while many writers were able to cast their minds into the future, only a few were able both to do so and to grasp the significance of looking back. Two who did were the astronomer Fred Hoyle and the techno-prophet Arthur C. Clarke, and both had assistance from historians in putting their ideas into deep context.

In his popular work of 1950, *The Nature of the Universe*, the astronomer Fred Hoyle wrote: 'once a photograph of the Earth, taken from outside, is available, we shall, in an emotional sense, acquire an additional dimension.' People had little concept of upward motion beyond aeroplane flight, thought Hoyle, but

> once let the sheer isolation of the Earth become plain to every man whatever his nationality or creed, and a new idea as powerful

as any in history will be let loose. And I think this not so distant development may well be for good, as it must increasingly have the effect of exposing the futility of nationalistic strife. It is in just such a way that the New Cosmology may come to affect the whole organization of society.

Hoyle went on to speculate about what a colour photograph of the whole Earth would look like.

There will be brilliant white patches where the Sun's light is reflected from clouds and snow-fields. The Arctic and Antarctic will on the whole appear brighter than the temperate zones and the tropics. There will be all shades of green, varying from the light green of young crops to the sombre darkness of the great northern forests. The deserts will show a dusky red, and the oceans will appear as huge areas that look grimly black, except occasionally they will be illuminated with a blinding flash where conditions allow the Sun's light to be powerfully reflected, much as we sometimes see a brilliant shaft of sunlight reflected from the windows of a distant house. The whole spectacle of the Earth would very likely appear to an interplanetary traveller as more magnificent than any of the other planets.

These passages had been composed early in the same year for a series of talks on BBC radio where, acknowledged Hoyle, 'many of the most graphic remarks and phrases were suggested to me by Mr Peter Laslett', who also insisted that they must be easily understandable. Laslett, a BBC talks producer with a brief to get academics to adapt their work for a radio audience, was also an up-and-coming historian; the astronomer's viewpoint was framed in the historian's perspective. Broadcast in the winter of 1950, the series achieved the highest audience appreciation rating of any series on the BBC's Third Programme; listeners pronounced themselves 'fascinated and awed' by the immensity of the concepts, presented with 'majestic simplicity' in Hoyle's friendly Yorkshire accent. The talks were repeated with similar success on the Home Service in the summer and went on to become a perennial seller for Penguin Books, providing the standard introduction to cosmology for a generation of British people.[4] After Apollo 12, Hoyle claimed that 'his earlier prediction had already come true: that once man had stepped off Earth, he would undergo a major revision in his self-image'.[5]

Hoyle later turned to science fiction writing to communicate his sense of progress to a wider public. The master of this technique was his contemporary Arthur C. Clarke. In the 1951 story 'If I forget thee, O Earth,' Clarke imagined a young lunar colonist gazing at the beautiful crescent of an Earth which he has never visited, and which he knows has been uninhabitable for centuries because of nuclear war: 'across a quarter of a million miles of space, the glow of dying atoms was still visible, perennial reminder of the ruined past.' In a similar spirit, Robert Heinlein's story 'The green hills of Earth' (1947) imagined the refrain of a travelling musician from the far future, singing to the settlers of some remote planet: 'We pray for one last landing on the globe that gave us birth, To rest our eyes on fleecy skies and the cool green hills of Earth.' These lines were broadcast to Dave Scott just before he returned to the lunar lander for the last time on Apollo 15: 'it helped to ease the pangs I felt as our spacecraft lifted away from the surface of the Moon,' wrote Scott.[6]

Later Clarke imagined the emotional power that the sight of Earthrise would have for an astronaut marooned in lunar orbit.

> The horizon ahead was no longer flat. Something more brilliant even than the blazing lunar landscape was lifting against the stars. As the capsule curved around the edge of the Moon, it was creating the only kind of Earthrise that was possible – a man-made one. In a minute it was all over, such was his speed in orbit. By that time the Earth had leaped clear of the horizon, and was climbing swiftly up to the sky. It was three-quarters full, and almost too bright to look upon. . . . The sight of the rising Earth brought home to him, with irresistible force, the duty he feared but could postpone no longer.

The astronaut's duty was to make a last call home.[7] For Clarke's stranded astronaut, as for the crew of Apollo 8, Earthrise came as a surprise.

In the 1950s Clarke was busy popularising the case for space travel. In *The Exploration of Space* (1951) he looked forward to the time when parochial mentalities would be replaced by 'a world outlook': 'few things will do more to accelerate that outcome than the conquest of space. It is not easy to see how the more extreme forms of nationalism can long survive when men have seen the Earth in its true perspective as a single small globe against the stars.' He may have had Hoyle's recent broadcast and book in mind, but in 1943, during the depths of the war, Clarke had speculated to the Christian writer C. S. Lewis about how 'national rivalries, which have caused most of the misery of the past, will finally appear in their proper perspective when they can be seen against the background of the stars'.[8]

After the war, Clarke developed his ideas after hearing a lecture given in 1946 at Senate House, University of London, by the historian Arnold Toynbee. Toynbee was then writing his multi-volume *History of the World*, a grandstand account of the rise and fall of civilisations, and his lecture was entitled 'The unification of the world'. Clarke was 'much taken' with it. 'It seemed to me,' he wrote, 'that we would be presented with a classic example of this when the space age opened. Here without question was the greatest physical challenge that life on this planet had faced since the distant days when it emerged from the sea and invaded that other hostile environment, the arid, sub-scorched land.'[9] Clarke went on to give a lecture to the British Interplanetary Society on 'The challenge of space', arguing: 'interplanetary travel is now the only form of "conquest and empire" compatible with civilization. Without it, the human mind, compelled to circle forever in its planetary goldfish bowl, must eventually stagnate.'

Both Clarke and Hoyle had been influenced by the humanist tradition of internationalist idealism which had come to maturity during the Second World War and after. So too had the American poet and Librarian of Congress, Archibald MacLeish. 'We know, all of us, the power of images in our lives and in the lives of nations,' MacLeish had mused as long ago as 1942.

> Never in all their history have men been able truly to conceive the world as one: a single sphere, a globe having the qualities of the globe, a round Earth in which all directions eventually meet, in which there is no center because every point, or none, is center – an equal Earth which all men occupy as equals. The airmen's

Earth, if free men make it, will be truly round: a globe in practice, not in theory.[10]

Here was the answer to Russell Schweickart's admiring question about MacLeish's instant meditation on the Apollo 8's view of the Earth: 'How did he know that?'

The global vision of peace was set forth in a book of essays on political geography, *The Compass of the World*, published in 1946. Its essays, many of which dated from the war years, dealt with subjects such as 'The myth of continents', 'The logic of the air', and 'The round world and the winning of the peace'. In 'The peaceful solution of boundary problems', the political geographer S. Whittemore Boggs imagined how invisible national boundaries would appear to a perplexed alien – not from space but 'if he were to visit a typical boundary and see how inconspicuous it is'. MacLeish's generation of idealists, however, had been intoxicated between the wars by the airman's vision of the world from above, free and open, at once modern and God-like.[11] In a swords-into-plowshares movement, long-range air travel with its potential for peaceful interchange was the product of the bomber technology of the war, just as space travel was to be the product of the intercontinental ballistic missile technology of the Cold War. Orthographic (globe-like) map projections by the innovative cartographer Richard E. Harrison – one of them entitled 'One world, one war' – emphasised how different the world looked from above. Viewed from the pole, north America and northern Russia were neighbours round the Arctic Circle; in the air age, the Arctic Ocean was the new Mediterranean.

The Nazi view of the world by contrast was a flat, two-dimensional one based on power alone. In Nazi cartography, great black arrows shot out from the German homeland to grab at the 'world island' of Eurasia and Africa, with the freedom-loving New World pressed back to the margins across great expanses of ocean. Victory for the Allies, proposed MacLeish, depended upon winning a battle of visions, between 'the Nazi image of the airman's Earth or ours', between a tyrannical ground-based domination and the freedom of the air: 'if those who have the mastery of the air are free men and imagine for themselves as free men what their world could be, their world will be the full completed globe – the final image men have moved toward for so long and never reached.'[12] In 1945 Germany surrendered not to Britain, or the USA, or even the Allies, but to the United Nations: to the world.

Utopianism flowers like the snowdrop in the darkest times, as Jay Winter has written; the 1940s were a case in point.[13] But while the Second World War had made global visions seem like practical politics, it had not created them. Between the wars, when the centuries-long age of expansion appeared to be coming to an end as imperial and totalitarian states struggled for living space on a fully colonised planet, the first generation of astrofuturists dreamed of escaping into space. Before the war, Clarke had felt sure that the future history of mankind lay in space, rather than 'cooped and crawling on the surface of this tiny Earth'.[14] Adlai Stevenson, a Democratic politician of the same generation, later put it like this in a speech as US Ambassador to the United Nations:

> Just as Europe could never again be the old closed-in community after the voyages of Columbus, we can never again be a squabbling band of nations before the awful majesty of outer space.
>
> We travel together, passengers on a little space ship, dependent on its vulnerable reserves of air and soil; preserved from annihilation only by the care, the work and I will say the love we give our fragile craft. We cannot maintain it half fortunate, half miserable, half confident, half despairing, half slave . . . half free. . . . No craft, no crew, can travel safely with such vast contradictions.[15]

For astrofuturists such as Clarke, the vision of Earth as a planet was associated with the urge to leave it; for liberals like Stevenson, it was associated with the urge to improve it.

The Russian writer Yuri Melvill offered a Soviet version of astrofuturism. 'The farther man's spaceships travel, the more distinctly will he see the Earth's beauty and unity,' he wrote in 1966. 'Seeing the Earth in celestial perspective will give him a new insight into the indivisibility of the human race, its common origin and common cosmic destiny.' But realising such a vision was beyond western capitalism, argued Melvill, for it required 'peace . . . the elimination of colonialism [and] an end to all kinds of oppression. . . . It is only "socialized mankind" – to use Marx's expression – that will be equal to the task.'[16] Melvill's article, in an English language journal intended for the West, may have been a response to Adlai Stevenson's speech at the UN; it was certainly a challenge to the assumption that western values were as universal as they claimed to be.

A figure who tried to stand on both sides of the divide was the astrofuturist and social reformer David Lasser, whose life story has been wonderfully recovered by De Witt Douglas Kilgore. In his 1931 book *The Conquest of Space*, Lasser argued that the sight of the Earth from space would break down racial divisions.

> A great spiritual tranquillity fills us – a humbleness and a yearning for the continuance of this immense peace. Our being seems spread through the eternity that we can see. We realize now the full meaning of Einstein's 'cosmic religion'. Cities, empires, states; dreams and ambitions; conflict and confusion are infinitely remote, part of the dream-world of that slowly-turning globe.

The reference was to Einstein's recent essay on 'Religion and science' in the *New York Times* magazine: 'the cosmic religious feeling is the strongest and noblest motive for scientific research,' he had written. In the end, although Lasser never lost his vision of a common human destiny in space, he felt obliged to set aside practical space advocacy in favour of a career as a trade unionist and government administrator: 'I decided that solving our earthly problems had to come first,' he explained to Arthur C. Clarke.[17]

No one felt the 'cosmic religious feeling' more powerfully than Olaf Stapledon, who provided what is still the grandest ever vision of cosmic evolution in his novels *Last and First Men* and *Star Maker*. His visionary yearnings for the 'crystal ecstasy' of the cosmos served as the link between the Victorian futures of Jules Verne and H. G. Wells and the hard science fiction of the mid-twentieth century. 'No other book had a greater influence on my life,' Clarke wrote of *Star Maker*, and he was not alone. Clarke invited Stapledon to give a widely publicised talk to the British Interplanetary Society in 1948.[18] Stapledon died relatively young in 1950; had he lived well into old age, he would have learnt just how prophetic was his vision of planet Earth.

In *Star Maker* (1937), Stapledon imagined himself 'soaring away from my home planet at incredible speed' as if in a dream, until the

Earth was like 'a broad disc of darkness surrounded by stars'. As the Sun rose, the Earth appeared as 'a visibly waxing Moon' until it was full and bright, 'an indefinite breadth of luminous haze, fading away into the surrounding blackness of space'. But while he had thought through the details, Stapledon's sense of the Earth was essentially a spiritual one. From space, there was no sign of human activity.

The spectacle before me was strangely moving. Personal anxiety was blotted out by wonder and admiration; for the sheer beauty of our planet surprised me. . . . It was far more lovely than any jewel. Its patterned colouring was more subtle, more ethereal. It displayed the delicacy and brilliance, the intricacy and harmony of a live thing. Strange that in my remoteness I seemed to feel, as never before, the vital presence of Earth as a creature alive.

Earth was also fragile, 'this whole grain of rock, with its film of ocean and of air, and its discontinuous, variegated, tremulous film of life'. Here, from a secular humanist yearning for something to worship, was a premonition of Gaia, the living Earth. A decade earlier, the Russian biologist Vladimir Vernadsky had foreseen that 'the face of the Earth viewed from celestial space presents a unique appearance, different from all other heavenly bodies', precisely because of its biosphere; Stapledon is very unlikely to have known this.[19]

The grandfather of all astrofuturists was the Russian writer Konstantin Tsiolkovsky, best known in the West for his description of the Earth as mankind's cradle; his extraordinary life awaits its first western biographer. Tsiolkovsky's didactic science fiction story *Beyond the Planet Earth* was first published in Russia in 1916, but begun in 1896. In it, the first voyage into space is undertaken in secret a hundred years hence by an international community of settlers led by a scientist named Ivanov. Having ascertained that they are indeed in orbit, he orders the window shutters to be opened. 'Some who were standing at other windows saw the Earth, thousands of kilometres away. At first they did not even know what they were looking at.' The Earth, Sun and stars seem close enough to touch. One traveller exclaims: 'How strange our Earth looks! It takes up nearly half the sky, and looks concave, like a bowl, instead of convex; as though it were inside this bowl that the people live.'

The brim of this bowl as seen by our travellers is very uneven, dotted here and there with mountain peaks which stand out like huge teeth. Away from the edges there is a haziness, and further still a quantity of oblong grey patches. These are clouds, darkened by a thick layer of atmosphere.

. . .

Some of the watchers were overcome or exhausted by the spectacle, or driven from their windows by it. Some, deterred by the exclamations of their friends, had not even looked out. Many flew into their cabins, closed the shutters and put out the feeble electric light. Others flew impatiently from one window to the next, feasting their eyes . . . like children in a railway carriage or on a steamer for the first time in their lives. What most drew their attention was the Earth.

They are fascinated by 'the gigantic sickle of the Earth' as it wanes, by the shadows of mountains jutting beyond the sunset line, by the way the dark part of the Earth is faintly lit by the Moon, and by the sight of space just beyond the dark rim of the Earth, where 'new stars were being born as if from nowhere'. The settlers debate whether the freedom of space is preferable to life on Earth with all its poetry; 'isn't man himself the highest poetry of all?' they ask themselves. Tsiolkovsky's travellers also view the Earth from near the Moon, but here his imagination fails him: 'from farther off, it looked no different, except that the scale was smaller.' As they head towards the asteroid belt, the Earth becomes 'more like a bright star than a planet'. After surveying the solar system for habitable zones they return to an Earth which, under a benign world government, is preparing for a mass migration into space.[20]

Before the mid-twentieth century, for most people the nearest approach to a vision of the whole Earth was to see one of the giant globes on display at international Expos in major cities across Europe and North America. The original Great Exhibition, in London in 1851, didn't include a globe, but a private showman, knowing the appeal of stage spectacles and panoramas, saw an opportunity. During the Great Exhibition period, visitors to Wyld's Great Globe in Leicester Square paid to view a giant globe with plaster-cast oceans and continents.[21] Over a century later another huge globe graced the 1964 World's Fair in New York, where Arthur C. Clarke and Stanley Kubrick went together to seek inspiration for their forthcoming collaboration on the film *2001: A Space Odyssey*.

The 1900 World's Fair in Paris had a particularly spectacular example of a globe, sitting next to the Eiffel Tower (albeit painted with animals rather than continents), and the exhibition itself was devoted to the idea that global trade would bring about world peace. Jay Winter has identified 1900 and the global projects it led to as one of six 'utopian moments' in the twentieth century. One of those inspired was the French banker and philanthropist Albert Kahn, who in the early twentieth century embarked on a project 'to photograph the whole world, and to preserve it in Paris for all to see as an archive of the Planet'. Kahn sent out teams of photographers to the most distant parts of all five inhabited continents to 'capture the face of humanity'. They returned with some 75,000 photographs and 300 miles of film, much in colour. An internationalist and a pacifist, 'Albert Kahn believed firmly that his archives of the planet would show us all what we had in common, thereby making war unthinkable'.[22] Kahn's methods were novel but we can now identify his belief, that the sight of the whole Earth would create a sense of shared destiny that would bring its people together, as the start of a long twentieth-century utopian tradition.

Such thoughts were far from the mind of H. G. Wells, who, as the new century opened, published one of the first science fiction novels, *The First Men in the Moon* (1901), which included an imaginary view of the whole Earth. His lunar voyagers travelled in a capsule propelled upwards by a mysterious anti-gravity process. Wells offered his travellers only a thirty-second glimpse of the Earth from 800 miles up:

> With a click the window flew open. I fell clumsily upon hands and face, and saw for a moment between my black extended fingers our mother Earth – a planet in a downward sky.
> . . .
> Already it was plain to see that the Earth was a globe. The land below us was in twilight and vague, but westward the vast grey stretches of the Atlantic shone like molten silver under the receding day. I think I recognised the cloud-dimmed coast-lines of France and Spain and the south of England.

The window was quickly closed to prevent further vertigo. Ironically, while some early twentieth-century thinkers were able to encompass the whole Earth, Wells the ultra-progressive humanist could only imagine it as from a great height. The distant Earth gave Kahn a sense of peace and Stapledon a sense of life; Wells just got vertigo.[23]

The other great nineteenth-century science fiction pioneer, Jules Verne, imagined a similar journey in 1870 in *Around the Moon*. Verne's pioneers travel (somewhat unrealistically) in a projectile shot from a cannon, but in other respects their journey has remarkable parallels with that of Apollo 8. Three travellers are launched eastward from Florida in December, in a capsule of almost identical weight reaching similar speeds; they reach the Moon in a similar period and orbit it several times without landing; and they splash down in the Pacific. Like the crew of Apollo 8, they are fascinated by the sight of the distant Earth, 'its delicate crescent suspended in the deep blackness of the sky. Its light, rendered blueish by the thickness of its atmosphere, seemed less intense than that of the lunar crescent.' They at first mistake it for the Moon, until they realise that the rest of the globe is faintly visible in reflected Moonlight: 'the terrestrial crescent seemed to curve more than half way round its disc – an illusion caused by its brightness.'[24] This makes the Earth seem alien, like the Moon, as a planet rather than as home.

The view from the past

Moving back in time from Verne's early science fiction novel *Around the Moon*, we cross the boundary to a much earlier, non-futuristic tradition of imagining the whole Earth. Verne's previous novel had been *Five Weeks in a Balloon*. A generation earlier, Edgar Allan Poe had told the tale of Hans Phaall, who ascended in a balloon all the way to the Moon. From seventeen miles up, 'the view of the Earth . . . was beautiful indeed' as the ocean turned an ever deeper shade of blue and began to appear convex. The prospect was chastening rather than enlightening.

> Of individual edifices not a trace could be discovered, and the proudest cities of mankind had utterly faded away from the face of the Earth. From the rock of Gibraltar, now dwindled into a

dim speck, the dark Mediterranean sea, dotted with shining islands as the heaven is dotted with stars . . . seemed to tumble headlong over the abyss of the horizon.

From further out still, 'no traces of land or water could be discovered, and the whole was clouded with variable spots, and belted with tropical and equatorial zones'. In the end, the Earth was 'like a huge, dull, copper shield, about two degrees in diameter . . . tipped on one of its edges with a crescent border of the most brilliant gold'. In another story, 'The landscape garden', Poe speculated whether, viewed from a great height, the disorder of nature might prove to have been 'set in array by God, the wide landscape-gardens of the hemispheres'.[25]

During the Enlightenment, to which Poe's gothic visions provided a dark coda, the vision of the Earth as a planet was associated with science and secularism. Probably the first panorama of the whole Earth, a hollow globe called the 'Temple to nature and reason', had been exhibited in revolutionary Paris, home of balloon launches, as early as 1793.[26] In 1791, in the midst of the French Revolution, the traveller and historian Constantine de Volney published *The Ruins, or Meditations on the Revolutions of Empires*. In it a traveller, wearied and troubled by viewing the ruins of an ancient and apparently greater civilisation than his own, and by the suffering and turmoil of his own age, falls into a reverie in which a phantom 'Genius' wafts him to the heavens. 'I perceived a scene altogether new. Under my feet, floating in the void, a globe like that of the Moon, but smaller and less luminous, presented to me one of its phases; and that phase had the aspect of a disk variegated with large spots, some white and nebulous, others brown, green or gray.' The Earth was like the Moon 'seen through the telescope during the observation of an eclipse'; the seas were brown and the deserts white. 'What! said I, is that the Earth – the habitation of man?'[27]

For Volney, the great extent and variety of past human civilisation made traditional Christianity appear short-sighted and dubious. The Anglo-American radical Thomas Paine went further and argued in *The Age of Reason* (1793) that the discovery of life on other worlds would shatter orthodox Christianity for good: 'he who thinks that he believes in both has thought but little of either.' For sceptics such as

Volney and Paine, when the Earth appeared in its true context, humanity would be humbled, not before God but before the cosmos.[28]

The preceding age of enlightenment and scientific revolution seems not to have produced any significant visions of the whole Earth, which was increasingly conceived of in mechanistic terms.[29] In eighteenth-century Britain, astronomy lectures were perennially popular among the middle classes, animated by mechanical orreries in which small brass or wooden spheres representing the Sun, Earth, Moon and planets orbited gracefully at the turn of a handle. By the late eighteenth century tiny wooden globes were cheaply available, and carried in the pockets of working men. Monarchs in the Baroque period invested heavily in gilded models and other depictions of the globe as a way of asserting their own status as world-class rulers, following the example of Louis XIV, the 'Sun king'.

Against such grandiose pretensions there was cultivated a more pious tradition of depicting the globe as an invitation to contemplate with humility the limits of earthly existence. Painters and engravers depicted the globe as a bauble or a bubble, an emblem of the vanity of the material world. The globe in the philosopher's study was thus a cousin of the attractive but delicately mouldering apple in the classic still life study. For Roman Catholics there was also the genre of the cordiform map, in which the meridians were curved to give the world the appearance of a heart-shaped cushion, evoking the devotional practice of contemplating the sacred heart.

The 1630s saw what David Cressy has described as 'Europe's first space program' – a group of books speculating about the nature of the Moon, couched in the form of imaginary journeys. The year 1638 in particular was 'England's lunar moment', although it had been touched off by Johann Kepler's 1634 account of an imaginary visit to the Moon, and over the longer term by the news of Galileo's telescopic observations of the Moon and planets. This genre of cosmography mixed theology with exploration, providing a relatively safe format for the discussion of theologically sensitive issues such as the existence of intelligent life elsewhere in the universe. In 1638 both Bishop Francis Godwin and the natural philosopher John Wilkins published accounts of imaginary ascents to the Moon. 'The further we went, the lesser the Globe of the Earth appeared unto us,' wrote Godwin. It seemed like 'a huge Mathematicall Globe, leasurely turned before me, wherein successively, all the Countries of our earthly world within the compass of 24 howers were represented to my sight. And this was all the means I had now to number the dayes, and take reckoning of time.' The continent of Africa appeared 'like unto a Peare that had a morsell bitten out upon one side of him', and the Atlantic Ocean like 'a great shining brightnesse'.[30]

John Wilkins also decided that the Earth would dazzle from a distance on account of the sunlight reflected off the oceans which covered two-thirds of its surface, appearing rather like the Moon but even brighter. In support of his thesis, he quoted two contemporary writers. The mathematician Carolus Malapertius had suggested that 'if wee were placed in the Moone, and from thence beheld this our Earth, it would appear unto us very bright, like one of the nobler Planets.' The view of the Louvain theologian Libertus Fromendus was similar: 'I believe that this globe of Earth would appeare like some great starre to any one, who should looke upon it from the Moone.' In this classically minded age, Wilkins's trump card was a passage from the third-century Greek philosopher Plotinus: 'if you did conceive your selfe to bee in some such high place, where you might discerne the whole globe of the Earth and water, when it was enlightened by the Sunnes rays, 'tis probable it would then appear to you in the same shape as the Moone doth now to us.' The Earth's light, he believed, was enough to warm the Moon and create seasons and therefore life.[31]

Kepler, too, predicted a bright Earth, with the water brighter than the land, for on Earth 'whenever one looks at surfaces of land and water placed next to each other, the land is always dark and the water shines'. But unlike Wilkins, who saw life on the Moon as related to life on Earth, Kepler was worried about the implications of extra-terrestrial life: 'then how can all things be for man's sake? How can we be the masters of God's handiwork?'[32] Later the French philosopher Adrien Auzout tackled the question of life on the Moon. He cleverly reversed the viewpoint, asking himself what signs of life might appear on the Earth when seen from space. The changing colours of the seasons, he reasoned, would be obvious, and human activity would also leave visible traces:

we cut down whole Forrests, and drain Marshes, of an extent large enough to cause a notable alteration: And men have made such works, as have produced changes great enough to be perceived. . . . And when Fire lights upon Forrests of great extent, or upon Towns it can hardly be doubted, but these Luminous Objects would appear either in an eclipse of the Earth, or when such parts of the Earth are not illuminated by the Sun. But yet I know no man, who hath observed such things in the Moon.

This line of thinking would be followed three centuries later by James Lovelock, who, in devising tests for the existence of life on Mars, asked himself how a life-bearing planet might be detected from space.[33]

These seventeenth-century visions of Earth as a planet comparable to the Moon depended upon a Copernican view of the solar system in which Earth was indeed just a planet like others going round the Sun. But pre-Copernican writers were also capable of imagining the whole Earth. The late sixteenth-century writer Luis de Camões imagined Vasco de Gama on a mountain peak in the realm of the goddess Tethys, contemplating the crystalline sphere of the cosmos, 'infinite, perfect, uniform, self-poised', with the Earth floating at the middle.[34] The master engraver Albrecht Dürer also approached a sense of the whole Earth in his 1513 print of 'the imaginary orb' of Earth, produced with the aid of the Nuremberg scientist Johann Stabius, which depicts a spherical map projection invented by the fourth-century Greek geographer Ptolemy.[35] A line drawing animated by Dürer's unusual spatial awareness, it provides one of the first three-dimensional representations of the Earth as a planet, as opposed to the figurative globe of the geographers.[36]

Nothing, however, comes close to the prophetic work of the Portuguese artist Francisco d'Olanda, who in the 1540s produced a series of paintings to illustrate the biblical Creation story. On the third day the curve of the Earth appears as the land and the waters are divided, and on the fourth day comes the creation of the Sun and Moon. There, near the Sun and Moon, floating in space, is the Earth. Unlike the blue and brown Earths depicted right up until the 1950s, d'Olanda's Earth is blue and white. Oddly enough, it is the land which is blue and the seas which are white, as was the cartographical convention of the time, but it is the first image of the Earth as a planet to get the colours approximately right until the coming of space photography. Accompanied by the words of the Book of Genesis, d'Olanda's vision of the Creation offers an uncanny premonition of the flight of Apollo 8 over four hundred years later.[37]

Contrary to popular myth, the world was not widely believed in the Middle Ages to be flat; acceptance of a spherical Earth was virtually unanimous, and some thinkers suggested that it was the Earth, rather than the heavens, which revolved. The impression given by medieval maps of the world is partly responsible for the flat Earth myth. The typical *mappa mundi* showed a large central Eurasian-African landmass, arranged around the Mediterranean with Jerusalem at the centre. Circling it was a great ocean of unknown extent, with magic at the margins: sea serpents, cherubim and angels. This flat appearance reflected the limited perspective techniques of the period, and the fact that no one knew how the world joined up at the back. In a fifteenth-century work, *L'Image du monde* by Gautier de Metz, there is a whimsical illumination showing the Earth as a globe, houses and trees sitting atop a light crescent of ocean, floating in a blue, star-filled sky.[38]

The medieval model of the cosmos could not admit the concept of space travel. In this model, portrayed in a diagram of 1493, each planet was fixed to a series of concentric crystalline spheres. The outermost sphere was in turn encased by that of the fixed stars. Within lay those of the five known planets, the Sun, and the Moon. At the centre lay the Earth, a captive mudball; somewhere inside was hell. The system had no life of its own; sitting outside, God and his angels had to keep it all moving. The heavenly bodies, moving in circles, were perfect and unchanging; only on Earth was there change, which was associated with imperfection, decay and death. Travel between the spheres was physically impossible; the people of Earth were fixed to its surface (or six feet under it) until the day of judgement. The Earth, in this model, was the poor relation of the other planets, cast down far from heaven; Aristotle's gloomy cosmology complemented the worldview of early Christianity. But if we keep travelling backwards in time, through the Dark Ages to the classical world, we start to encounter visions of the whole Earth once again.

In classical philosophy, imagining the Earth was a philosophical tool to assist humble contemplation of the status of mankind. In the first century BC Cicero has the late Roman general Scipio Africanus appear in a dream to his grandson, also called Scipio, to ascend with him beyond the Earth. 'Men were created with the understanding that they were to look after that sphere called Earth,' explains his grandfather, pointing out the desert belts and polar regions. 'The inhabited portions on Earth are widely separated and narrow, and . . . vast wastes lie between these inhabited spots, as we might call them; the Earth's inhabitants are so cut off that there can be no communication among different groups.' He directs Scipio's attention away from the Earth to the rest of the universe, urging, 'look closely, at how small is the portion allotted to you!' 'I was amazed at these wonders,' confesses Scipio, 'but nevertheless I kept turning my eyes back to Earth. . . . The Earth seemed to me so small that I was scornful of our empire, which covers only a single point, as it were, upon its surface.'[39]

Two hundred years later the same tradition was continued by the Syrian writer Lucian, who related the fable of 'Icaromenippus, the sky-man', who flies to the Moon using an eagle's wing and sees it with an eagle's eye. 'The Earth you see is very small, far less than the Moon,' explains Menippus on his return; he wouldn't have recognised it but for landmarks such as the Colossus of Rhodes, and the way 'the ocean glinted in the Sun'. From afar, 'the life of man in its entirety disclosed itself to me . . . all the life that the good green Earth supports'. Like Scipio, he is impressed by the smallness and folly of human life:

> I was especially inclined to laugh at the people who quarrelled about boundary lines . . . for it seemed to me that the widest-acred of them all had but a single Epicurean atom under cultivation. And when I looked toward the Peloponnese and caught sight of Cynuria, I noted what a tiny region, no bigger in any way than an Egyptian bean, had caused so many Argives and Spartans to fall in a single day. . . . The cities with their population resembled nothing so much as ant-hills.

The Roman historian Seneca had similar thoughts on a vision of the Earth: 'Is this that pinpoint which is divided by fire and sword among so many nations? How ridiculous are the boundaries set by mortals.'[40]

The Roman poet Ovid, around the time of Christ, imagined Earth 'poised in the enveloping air, balanced there by its own weight'.

The earliest whole Earth vision of all, however, is that of Plato in the fourth century BC:

> The true Earth, if one views it from above, it is said to look like those twelve piece leather balls, variegated, a patchwork of colours of which our colours here are, as it were, samples that painters use. The whole Earth is of such colours, indeed of colours far brighter still and purer than those: one portion is purple, marvellous for its beauty; another is golden, and all that is white is whiter than chalk or snow; and all the Earth is composed of other colours likewise, indeed of colours more numerous and beautiful than any we have seen. Even its very hollows, full as they are of water and air, give an appearance of colour, gleaming among the variety of other colours, so that its general appearance is one of continuous multi-coloured surface.[41]

It was a remarkable anticipation of the bright, colourful and abstract globe which would be revealed by the early space photographs.

Seneca's thoughts about the Earth as a pinpoint and the absurdity of 'the boundaries set by mortals' were widely circulated in the sixteenth and seventeenth centuries as part of the caption to Ortelius's world map, the standard atlas of the Renaissance. In this and in other ways, the ancient vision of the Earth nourished a western cultural tradition of humble contemplation of man's place in the universe. Twentieth-century commentators on the first whole Earth photographs in turn unconsciously replicated these ancient thoughts. 'There is no better technological demonstration of the folly of human conceits than the distant image of our tiny world,' wrote Carl Sagan. The astronaut Harrison Schmitt, who took the famous 'Blue marble' photograph in 1972, said: 'now, so far from childhood's home, we see the planet revolve beneath us. All the works of man's earlier greatness and folly are displayed in our window in the course of a single day.'[42]

When Russell Schweickart paused in a spacewalk on Apollo 9 to gaze down on an Earth with 'no frames, no boundaries', he was part of a tradition almost as old as literature itself. In the ancient world the vision of the whole Earth was associated with pessimism, in the early modern world with pietism, and in the twentieth century with peace, but until the twentieth century it had never stood for progress. In 1968 men at the bow wave of progress voyaged out and saw for themselves the vision of the ancients, an Earth apparently untouched by the human race. It was fitting that they searched within for meaning and chose to read not from some manifesto of human progress but to start again: 'In the beginning . . .'

CHAPTER 4

From landscape to planet

'So Columbus was right!' was one comment when the first widely seen pictures of the curvature of the Earth appeared in 1948. Before people could see the whole Earth, they could see that it was curved. In a way, this might have been equally significant. But when exactly was the first photograph to show the curvature of the Earth? Strange to say, there has been no serious attempt to answer that simple question, perhaps because it turns out to predate the space age. Scattered through the middle third of the twentieth century, in specialist archives, in forgotten technical articles and manuals, in long-unread issues of magazines, and in hoarded press releases and newspaper clippings, lies the scattered evidence for one of the untold stories of the twentieth century: the transformation of the Earth from landscape to planet.

The curving horizon

The earliest aerial photographs had been taken in the mid-nineteenth century from balloons, producing bird's eye views of Second Empire Paris and pre-civil war Boston. In the 1890s, experiments began with cameras carried on kites and rockets. As early as 1891 Ludwig Rohrmann was granted a German patent for a rocket to photograph Earth. After launch, the rocket would quickly take its pictures and then explode, throwing clear a parachute and camera attached to a long cable which was anchored to the launch site and would then be winched back. There is no record that it was ever tried out. Another unlikely patent, for a camera attached to a homing pigeon, succeeded in producing the original bird's eye view of the world; at this stage, homing pigeons were more predictable than rockets.[1]

In Sweden, Alfred Nobel (founder of the prizes) launched the first photo-rocket over a small town in 1897, and a few years later Alfred Maul photographed a stretch of German landscape from a photo-rocket several hundred feet up. The Chicago-based photographer George Lawrence, whose slogan was 'the hitherto impossible in photography is our speciality', made his name with a photograph of 'San Francisco in ruins' after the 1906 earthquake, taken with a large camera suspended from an array of seventeen kites.[2] In terms of altitude, however, none of these contraptions offered an improvement on the view from the top of a hill.

Aeroplanes produced more elevated bird's eye views of the world, exciting the western imagination to a range of futurist visions, but the horizon remained obstinately flat.[3] In the 1930s high altitude balloons were at the cutting edge of the air age, easily exceeding the heights achieved by aeroplanes. In theory it might have been possible to detect the curve of the horizon from a plane flying a few miles up, given a clear view over plain or ocean and well-calibrated viewing equipment, but in practice atmospheric haze obscures the very small amount of curve at such altitudes. Before the horizon would start to curve it was necessary to get above the haze layer and into the stratosphere, and that still required a balloon.

In the late 1920s and early 1930s balloons repeatedly broke the altitude record, rising to over 60,000 feet (11 miles, or 18 kilometres). Elements of the altitude race foreshadowed the space race. The Russians claimed two record-breaking missions, but one could not be substantiated and the other ended in a crash that killed the occupants. The US Army and Navy challenged each other for the lead almost as vigorously as they challenged the Russians. With the Navy holding the official record, the US Army Air Corps decided in 1934 to team up with the National Geographic Society for a record attempt, named Explorer. To launch such a large balloon required exceptionally calm conditions; even Arizona's Meteor Crater wasn't quite right. After much searching they found a large natural bowl in the Black Hills of South Dakota, locally known as Moonlight Valley but now renamed the Stratobowl. The first attempt, in July 1934, was just short of the record height when the balloon developed a massive tear. There followed a sort of slow free fall of 11 miles as the crew struggled to guide the ruined balloon like a giant parachute.

The cameras were destroyed in the crash but the crew baled out near the ground.[4] Two of the three crew, Captain Albert Stevens and Orvil Anderson, undertook a second mission a year later, this time obtaining a historic photograph of the curvature of the Earth.

On 11 November 1935 Explorer II floated free of the Stratobowl. Eight hours later it was hanging in the stratosphere 13.7 miles (22 kilometres) high. Only 4 per cent of the Earth's atmosphere remained above it. The sky was nearly black, the Sun white, and with no air to bite on the balloon hung motionless. From behind the portholes of their pressurised gondola the aeronauts reported a feeling of 'profound calm'. They could see about 175 miles into a band of white haze that obscured the far horizon. Like the Apollo astronauts viewing the Moon, they saw the landscape below as 'a foreign and lifeless world'. Their large Fairchild vertical aerial camera, fixed to the floor, photographed the landscape down-wards. But they also took a hand-held Fairchild camera, loaded with infrared film, fixed in the gondola wall looking out across the landscape.

From the stratosphere the crew conducted a radio interview. 'The sky appears very dark indeed, but it can still be called blue . . . a very dark blue,' they reported. The announcer explained: 'The men antic-ipated and hoped that one of the cameras would be able to take a picture of the horizon which would show the curvature of the Earth. The buzzing you hear in the background is . . . one of the cameras automatically working.' It could see through the atmospheric haze to 330 miles (530 kilometres), and produced the first photograph clearly to show the curvature of the Earth. The enlarged photograph was published in a fold-out supplement to *National Geographic* a few months later. The field of view easily covered the whole of the Black Hills and its extensive river systems, extending to Wyoming and Montana; in the far distance, all detail was lost. A black line was ruled just under the horizon: above it could be seen, for the first time, the gentle curve of planet Earth.[5]

The achievement of Explorer II was certainly remarkable, but to take photographs of really big chunks of the Earth would require a rocket. Seriously useful rockets arrived in the United States in the form of a consignment of captured German V-2s – Hitler's 'Vengeance' weapons – shipped over after the Second World War.

Rocket men

Firing of the V-2s from the Army's Proving Ground at White Sands, New Mexico, began in the summer of 1946 and marked the start of the American space programme. 'The V-2 looked awesome, even empty, tame and unarmed,' recalled one scientist.[6] White Sands was also the site for upper atmospheric research, but unlike the smaller sounding rockets the massive V-2 could carry plenty of equipment. Packages of instruments weighing a ton or more were placed on board to analyse the upper atmosphere: 'Cosmic rays mystery pierced with aid of Nazi rocket' was one tactless headline in the *Washington Post.* Cameras were sent up as well, not to admire the view but to gather data on what was happening to the rocket. On V-2 shot number 12 on 10 October 1946 the Naval Research Laboratory physicist Thor Bergstrahl used a K-25 large-format aircraft camera to photograph the rocket's fins against the background of the Earth in order to work out how it was oriented. It worked, but the images were blurred and it was six months before the NRL team got another shot. The next chance to photograph the curvature of the Earth fell to another set of researchers: the team from the Applied Physics Laboratory (APL) at Johns Hopkins University, newly founded to exploit the scientific opportunities of the V-2 tests.[7]

The man responsible for putting cameras on the V-2s was APL's Clyde T. Holliday, a photographic specialist who during the war and after had worked on missile tracking and aerial reconnaissance. Getting photographs from a missile was no easy task, as Bergstrahl had already found: a camera with sensitive film, springs and a mechanical shutter had to survive a journey on a weapon of war. Under the New Mexico Sun it was packed next to liquid oxygen tanks at minus 183 degrees centigrade, violently vibrated on take-off, accelerated to twenty miles high in just over a minute, frozen in the upper atmosphere, spun with the rocket as it ascended, tumbled over and over as it descended, and finally smashed into the ground from a height of over fifty miles. Through all this it had to keep clicking and winding on automatically, with each click timed and electronically reported to the ground in order to work out exactly what was happening to the rocket.

Holliday experimented with various types of camera, including the trusty K-25 aircraft camera, a smaller 16mm gunsight aiming camera, and a commercial De Vry 35mm movie newsreel camera. All were strengthened to withstand vibration, with the film wound into hardened steel canisters. The main problem, though, was retrieving the film intact after the crash. Instrument packages at White Sands rocket programmes had been variously ejected, with parachutes, in slow-falling non-aerodynamic modules and even with rotor blades added; smaller packages drifted a long way and could take days or even weeks to find in the desert, even with the aid of smoke canisters, coloured ribbons and beacons. The V-2 solution was simpler. Beneath the rocket's shell were two spherical fuel tanks; in between was the camera. If these were blown apart on the way down, the lighter rear section, including the camera, would fall more slowly. The solution to finding the camera itself was even simpler: paint it red. There was still a lot of driving around the desert and grubbing in craters, but not a single V-2 film canister was lost.[8]

The thirteenth V-2 flight on 24 October 1946 carried a 35mm movie camera. The film survives; the whole trip lasts just a few minutes. The Explorer balloon had taken eight hours to reach a height where the horizon began to curve; the V-2 gets there in seconds. After just over a minute its fuel is exhausted. As the engines cut off and the rocket's fins fail to find enough grip on the thin atmosphere, it begins to spin, and the camera spins with it. As the V-2 slows near its apogee, some sixty-five miles up, it begins to tumble end over end. Sections of Earthscape sweep by at crazy angles, each sunlit landscape bounded by a curve of preternaturally dark sky.[9] 'The most sensational newsreel pictures of all time,' exclaimed one newsreel service when the pictures were released. 'You're on a V-2 rocket 65 miles up!' enthused the *Los Angeles Examiner*. The press loved the photographs but the director of the Science Service, like the nosecone-fixated engineers of the early space programme, just didn't see the point. The 'really important photograph will be a Sun – not the Earth,' he complained to his staff: did they get any pictures of the Sun? The meteorologists weren't impressed either. Why did the rockets always have to be launched on a clear day, they asked; where were all the interesting pictures of clouds?[10]

The first hundred mile high picture (160 kilometres) was taken by Bergstrahl's Naval Research Laboratory team in March 1947; several frames were patched together to make a mosaic landscape comparable to the early orbital pictures, stretching across Mexico and southern California to a horizon in the Pacific some 900 miles (1,400 km) away. Holliday's APL team capped this achievement in 1948 with the most spectacular V-2 picture of all. On 26 July 1948 Holliday collected over 200 photographs taken by an automatic camera on a V-2 some sixty miles up. A separate set of photographs was taken from a Navy Aerobee seventy miles up just over an hour later. After three months' work matching and stitching, two dramatic panoramas were released on 19 October. The view from the V-2 extended across 2,700 miles of horizon looking out over the Gulf of California, from Wyoming in the north to Mexico in the south. Extending over a tenth of the Earth's circumference, the landscape appeared as a shallow dome bulging out into space. A second north–south strip from the Aerobee was released at the same time. APL collected together the highlights of international coverage in over a thousand newspapers and magazines under the title 'Columbus was right!', alluding to the myth that Columbus had set sail to prove that the Earth was round. The more obviously panoramic V-2 picture was accepted in the press and the archives as 'man's first view of the curvature of the Earth', an official position it has held ever since.[11] It was a view of America that Hitler had aspired to but never lived to see.

Atmospheric testing rockets – known as 'sounding rockets' – continued to be launched from White Sands throughout the 1950s.[12] Infrared sensitive black-and-white film had always been preferred for technical clarity, but in October 1954 a small colour movie camera was mounted on an Aerobee sounding rocket. The result was a colour montage made up of 116 images covering two-thirds of the width of the north American continent from Nebraska to the Pacific – the nearest thing yet to a colour photo of the Earth.

The highest pre-orbital pictures of Earth, however, were taken by the US Air Force, which in August 1959 managed to get photos from automatically timed cameras packed into the nosecones of Atlas and Thor missiles at heights of up to 800 miles (1,100–1,300 kilometres). This was well past orbital height, and these tests involved separate re-entry vehicles and a splashdown in the Atlantic Ocean. Black-and-white film was used as colour was not expected to survive contact with sea water, but on one flight an experimental strip of colour film was taken. The colours were (understandably) a bit washed out, but it worked. The wide-angle lens took in a vast area of north and south America and the Atlantic Ocean, extending to west Africa and almost to Britain, one-sixth of the Earth's surface in all. The General Electric Company, the contractor, apparently unaware of previous achievements, proclaimed the pictures as the historic first views of 'the image of the Earth from beyond the atmosphere'.[13] As technology reached ever higher, the Earth shrank.

The attention would shortly switch to satellites, but rocket photography had one more moment of glory in the form of the astonishing X-15 rocket planes. The achievements of these fastest of all piloted vehicles rivalled those of the first sub-orbital Mercury spaceflights. The X-15 was carried up to eight miles high by a B-52 bomber and then released, at which point the pilot would ignite a rocket engine which erupted like a firework for just over a minute. Once it cut out he would soar, weightless, in a vast parabola under a black sky twinkling with stars, his plane now a ballistic missile stabilised (like a space capsule) by small jets, before re-entering the atmosphere and landing under normal controls. The journey from somewhere over Mud Lake, Nevada to Edwards Air Force Base near the Pacific coast took eleven minutes. The planes flew at up to six times the speed of sound (4,000 mph, 6,400 kph) and at heights of up to 67 miles (103 kilometres). Those who flew above fifty miles were given astronauts' wings; enjoying the freedom of the stratosphere, they looked down on the Mercury astronauts trapped in their tin cans. Among the test pilots with 'the right stuff' was Neil Armstrong, the first man to walk on the Moon. Flying between 1959 and 1964, the X-15s carried cameras, capturing superb photographs of the curving Earth as they descended; there would be nothing quite like it until the space shuttle.[14]

While rocket pictures of the curving Earth occasionally hit the headlines, they do not seem to have produced any enduring awareness of the Earth as a planet. Their captions drew attention to familiar geographical landmarks. They showed people what they already knew: that the Earth was round. They mostly appeared in a slightly alien, infrared light, with dark seas and black skies – impressive, but not really like home. The claims of the rival services for each new photograph of the curved horizon – Explorer II, V-2 and Atlas – reflects this lack of impact, as well as a ban on information sharing. As so often happens with pictures of America, these stories briefly had the status of world news in the US but made less of an impression elsewhere. For the photographic historian Beaumont Newhall, 'the photographs recovered from these flights were remarkable, for they showed the rotundity as well as the curvature of the Earth; as conventional aerial photographs resemble maps, these resemble sections of a globe.'[15] We might call them not landscapes but Earthscapes. Contemplating them, Thor Bergstrahl mused: 'if we had satellites, it would be fantastic'.[16]

Eye in the sky

The Atlas photograph of the Earth from 800 miles up proved to be the last hurrah of rocket photography. Already the first satellite, Sputnik I, had been launched, in October 1957. It circled the Earth, emitting bleeps but perceiving nothing. The early American Explorer satellites, rushed into orbit to salvage American pride, were too small to carry optical equipment. The space age was nearly two years old before the first camera satellite, Explorer VI, was launched in August 1959 – the same month, in fact, as the Atlas photos. It orbited the Earth at 17,000 miles, easily high enough to capture the entire Earth disc in a single shot. While it had no camera, Explorer VI did carry 'a 2½ lb scanning device – something like a TV camera – which is designed to relay a crude picture of the Earth's cloud cover'. Its signals were processed by 'a tiny electronic brain . . . called Telebit'. The resulting 7,000 pixel image of a crescent Earth took forty minutes to transmit. The scanner had been optimised to study the Van Allen radiation belts, so several weeks of processing and enhancement ensued before the picture was deemed fit to be seen. Although it was 'far and away the remotest picture ever made of the Earth', it was crude compared with the recent Atlas missile photographs. It was revealed to the world on 28 September with an outline map superimposed to show that the Pacific Ocean and Hawaii were down there somewhere. 'Scientists can discern cloud banks in the large white areas,' declared NASA.[17]

Less than a month later the Soviets showed what could really be done when the Luna-3 probe transmitted a photograph of the entire far side of the Moon. The Americans, secretly monitoring the Soviet transmissions, saw the picture even before it was released, and were able privately to authenticate the Soviet claim.[18] The obvious question to ask was: if they can do that for the Moon, why not for the Earth?

The lack of any real visual evidence of America's early presence in space now seems a curious oversight, given the experience which had been built up. An American picture of the whole Earth from space, photographed from an open sky and published for the world to see, would have made a fine advertisement and helped to diminish the sense of inferiority in space. As it was, Kennedy was able to alarm voters during his 1960 presidential campaign with the information that 'the first photograph of the far side of the Moon was made with a Soviet camera'.[19] The space programme remained grimly focused on technological one-upmanship, soon to be hard-wired into the programme by President Kennedy's decision to aim for the Moon.

The earliest satellite photography of Earth was not undertaken directly by NASA but by the US Weather Bureau. Weather scientists had been fascinated by the possibilities of high-altitude photography of clouds ever since the early V-2 pictures. Harry Wexler, the US Weather Bureau's observer on the V-2 panel, kept his superiors well supplied with pictures of the Earth for their office walls.[20] There was one problem: the V-2s' frustrating habit of launching only on clear days. In partnership with NASA, the Meteorological Office launched the TIROS I weather satellite on 1 April 1960, hoping for bad weather. TIROS (which stood for Television Infra-Red Observation Satellite) orbited 450 miles (700 kilometres) up – much closer than Explorer. Its aim, meteorologists explained to the press, was to take 'pictures of clouds'. The TIROS pictures showed dramatically curved images of the globe, but this was a distortion created by the wide-angle lens.

TIROS may have had technical limitations, but it did have one important advantage over later satellites: it took oblique photographs, panoramas, rather than the vertical mapping-style shots familiar from later orbiters. TIROS I was in a polar orbit, synchronised with the Sun, so that like the Sun it passed over the same points on the Earth at the same time each day. It stabilised itself by spinning, with the camera pointing along the axis of spin. It was also orientated with respect to the stars, so that its angle relative to the Earth was always changing (imagine a spider dangling by a thread from the hand of a clock, always looking down but continuously changing its angle with respect to the clock face). This produced an interesting variety of angled panoramas. There were still security considerations about high-altitude photographs, so NASA had to wait nervously while a selection of images was cleared by the CIA. 'I'm not sure that we deliberately gave them a bad batch of pictures,' recalled a NASA official, but 'I wouldn't put it beyond us because I wasn't very sympathetic. . . . That was the end of the classification.'[21] In 1961 the Department of Commerce offered whole batches for sale at $4 a reel. Later, NASA assembled a long strip photograph across the whole diameter of the Earth from a mosaic of individual TIROS photographs, but it was buried in a technical publication and accompanied only by an extensive discussion of cloud patterns.[22]

TIROS even tried to take a photo of the Moon. Unfortunately, the resolution from a quarter of a million miles was just not up to it. 'We have tried,' explained a weary meteorologist, but 'the Moon itself is only two television lines wide when viewed, and finding among the noise spots the little spot that might have been the Moon has not been successful'.[23] Taking a good photograph of the whole Earth from distant space was not going to be easy.

The human factor

From its foundation in 1958, NASA was focused on the Moon, not on the Earth. At programme level (Mercury, Apollo and so on), decision-making was dominated by engineers and mission planners with a decidedly limited tolerance for 'tourist photographs'. A small advisory group on photography had been set up early on, reporting to head office in Washington, but it sat outside the management structures for the individual programmes so its influence over mission priorities was limited.[24] The task of championing astronaut photography as part of the manned space programme fell in practice to one man, Richard Underwood, who eventually became Apollo director of photography and was responsible for planning all the famous Apollo Earth photographs.[25]

Underwood had the ideal technical background for a pioneer of space photography. He had served in the Pacific fleet in the Second World War and afterwards volunteered as a guinea pig observer for the Bikini atomic bomb tests. Cameras were banned but Underwood had read Leonardo da Vinci and knew how to build a camera obscura, a type of pinhole camera used by artists to project images onto paper. Adding some photographic film which his mother sent hidden in the wrapping of a chocolate bar, Underwood managed to get two successful pictures of the bombs – 'a miracle of photography', according to the friend who processed them. The danger money for the atomic tests (he got seven months' double pay) financed a degree in engineering and geology at the University of Connecticut, where he learnt aerial photography from a light aircraft, 'hanging out of the door with a rope holding me in'.

Underwood went on to work for the US Army Corps of Engineers as a worldwide aerial photographer, operating a mapping camera while next to him a spy camera would be gathering more sensitive information. His visual memory and meticulous record-keeping became legendary. As the Apollo astronaut Walter Cunningham said, 'Dick Underwood is absolutely amazing. After taking pictures from all round the world, you can lay any one down in front of him and he'll know where you are.' He ended up with NASA after he lost his security clearance, having married a woman from Honduras. His chief engineer knew Wernher von Braun, whose rocket programme was now testing at Cape Canaveral and who needed a photographer. Von Braun called him and asked if he knew about aerial photography. Yes, replied Underwood, from 80,000 feet. 'We work at half a million feet,' said von Braun. Underwood took the job.

The best cameras to put on von Braun's Redstone rockets, decided Underwood, were fighter plane gun aiming cameras, which could already withstand extreme vibrations and G-forces. They laid out a test range over the Atlantic using the same methods as those used for aiming nuclear missiles over the poles towards the USSR. The photographs, however, were disappointing: 'with the Redstones, when we shot them out of the Cape we shot them down range. All we were getting pictures of was water.' All the same, the pictures of the curving Earth got Underwood thinking, 'because nothing had shown anything like that before, and the next stage up would be the total disc . . . I would have been thrilled to look at it.' In 1962 he was transferred to the Photographic Services Division at the new Manned Spaceflight Center at Houston, home of Mission Control. There, he began planning to take the ultimate photograph: of the Earth itself.

While immense resources were put into taking photographs *of* the first manned Mercury flight in May 1961, very little went into taking photographs *from* it. A hundred and thirty cameramen from RCA were employed over several months round the clock photographing every detail of the Mercury preparations, and on the flight itself both pilot and instruments were filmed continuously. The view from the capsule, however, was captured only by a remote-controlled wide-angle camera mounted in a periscope. It took photographs of 'sky, clouds and ocean', but these didn't really bring out the 'beautiful Earth' that Shepard saw.[26] On the second Mercury flight Virgil Grissom took over 300 shots, mostly duds. Twenty photographs, though, provided a continental-scale panorama of west Africa from Morocco to Chad, taken from a hundred miles up and looking across to a horizon 900 miles away. Unlike the usual ill-fitting aerial camera mosaics, they showed the landscape in perspective, all at one time, with one set of shadows, revealing hitherto undetected geographical features.[27] The V-2s had shown America from space; Mercury showed the world.

Allowing astronauts to take their own photographs was a different matter. Tom Wolfe in *The Right Stuff* has told the story of how the famous 'Mercury Seven' fought behind the scenes to be recognised as 'test pilots in space' rather than just human components – a 'man in a can'. They also had to fight to be allowed to take pictures. 'There was a great deal of objection to carrying cameras,' recalled Underwood, 'because cameras weighed and cameras took up space.' At first the Mercury capsules were not even going to have a window. John Glenn, however, insisted on both a large viewing window and a camera for his orbital mission of February 1962. There was no camera on the official equipment list so Glenn had to go right up to the Mercury programme director Robert Gilruth in order to be allowed to take one as part of his personal kit. After trying out several, the one that could be worked best with one gloved hand was an ordinary $20 Ansco camera he had bought in a drugstore on Florida's Cocoa Beach. It was modified for him by Underwood and his team. As he travelled three times round the Earth from west to east, facing backwards, it felt not that different from being in the cockpit of a rocket plane or a training simulator: 'there was no sense of taking in the whole Earth at a glance.' He took a picture of the Atlas Mountains curving across the horizon, and several more of the Sun setting, seen from beyond the atmosphere. The quality was weak but they showed what was possible: 'when we saw them, they were great!' remembered Underwood.[28] Neither Glenn nor any of the other first four Mercury astronauts received any serious training in photography. The fifth and sixth pilots, Walter Schirra and Gordon Cooper, received a joint three-hour briefing session on what was called the 'synoptic terrain photography experiment', to catalogue 'physiographic features of the Earth' and 'cloud patterns'. Schirra explained: 'my initial interest in the camera was not high . . . but when we had to go with a camera, at least I was sold on the idea of having to go with it, I decided we should go first class.' Rejecting 'dinky cameras' he asked to take his own professional quality 70mm Hasselblad, but was told no: it wouldn't work properly in zero gravity, the leather case would produce gases that could fog the window, and he could be blinded by unfiltered sunlight reflecting off the metal shell. He reacted 'like a three-year-old kid who broke his favourite toy', recalled Underwood, 'so it was determined that we'd modify one.' 'All the leather was taken off that beautiful Hasselblad,' recalled Schirra. Schirra's plans for photography also ran into problems stemming from the previous Mercury flight, when Scott Carpenter had been so captivated by the sight of Earth from orbit that he used precious fuel manoeuvring for a better view and landed 200 miles short of target zone. For the next mission, explained the official history, 'a prescribed ground rule of the flight was to conserve control fuel, [so] only a few selected photographs were taken'. In orbit, Schirra found himself 'depressed about the tremendous quantity of cloud coverage'. Struggling to wrestle the camera out of its bag, he took the prescribed pictures and then announced: 'I will take one of the horizon just for posterity.' Sadly, his fourteen exposures were mostly cloudy or overexposed.[29]

The last Mercury astronaut, Gordon Cooper, was the most experienced photographer of the group, and the pictures that he brought back from his twenty-two orbits were described by NASA's press officer as 'almost magazine quality'; they included some of the Himalayas and western China. Cooper claimed from orbit that he could see trucks, trains and smoke from buildings, observations which were first dismissed as impossible; it was later found that Cooper had exceptional eyesight, assisted by the reduced atmospheric blurring from space, and the fact that human vision was more sensitive to linear features than previously thought. The human eye did have a special role in space after all.[30]

There was a gap of nearly two years in the American manned space programme before the next phase, project Gemini, in 1965–6. The two- and three-man Gemini missions performed stately orbital manoeuvres, dockings and even spacewalks in rehearsal for the journey to the Moon. These generally lasted several days, giving the crew much longer to look out of the window. The main visual discoveries of the Gemini programme were summarised at the time: 'sky in space is black, the world has an unexpectedly high degree of cloud cover, and the Earth is a blue planet'.[31] A pivotal moment in awareness of the Earth came with the second Gemini mission, Gemini 4, in June 1965. The Russian cosmonaut Alexei Leonov had performed the first 'spacewalk' in March. Those present at the pre-launch press conference for Gemini 4 were intrigued by the hint that one of the astronauts might 'stick his head out of the hatch'. Ed White in fact spent twenty minutes floating outside the capsule. As far as the sense of speed or falling was concerned, he later recalled, it was 'similar to flying over the Earth from about 20,000 feet'. There were two important differences: he was 135 miles up, and there was no window between him and the Earth. White had a camera attached to the small gas gun which he used to manoeuvre himself, and with no window in the way he was able to take the clearest pictures yet seen of the Earth from space.[32]

After splashdown the pictures were rushed by air to NASA's Manned Spaceflight Center in Houston, developed overnight and laid out on tables for the flight team to inspect. The official purpose of White's camera had been to photograph the outside of the space-craft itself to check for any damage to the surface, and this was what interested Robert Gilruth, the director of the Manned Spaceflight Center. Underwood, however, was drawn to the pictures of the Earth, particularly one of the Nile delta.

Dr. Gilruth says to me . . . 'Hey, Dick, all the action's up here.' I says, 'Well, I don't think so, Dr. Gilruth. I think it's all down here.' And he came down and he looked and he said, 'Well, those are just pictures of the Earth.' I said, 'Yeah, but we're looking at things that no human being had ever seen before, parts of Africa and other places. You see what really goes on.' I explained all these things to him. He said, 'Well, from now on your job is to work with the astronauts to be sure they bring back great photos of the Earth and then eventually we go to the Moon.' . . . That was the pivotal point.[33]

But while senior NASA people understood the value of Earth photography, Underwood still found himself pushing uphill:

I was the only supervisory aerospace technologist in that whole directorate, and that made a lot of people unhappy. They kept trying to fire one supervisory aerospace technologist in the directory and I was the only one. But Gilruth and Low and other people were on my side, so it never worked out. . . . Mr Webb and his successors thought space photography was great.

The engineers who complained that photography 'wasn't technical enough' were persuaded by experience that it could be useful for documenting things that went wrong. Even so, recalls Underwood,

I used to scream at the engineers in meetings . . . 'You're going to spend $50 billion on everything else, yet you don't want to spend $20,000 on cameras . . . without those pictures, we'll have no idea of what happened up there. You can load thousands of books with all this computer data about our trips to the Moon, shove them in a library, and nobody will ever read one.'[34]

But behind all the technical justifications for photography, his main aim was simple: 'to get these great Earth-looking pictures'. Brian Duff was at NASA's Public Affairs Office in Houston:

My recollection of those days is a constant battle with my friend Deke Slayton and the other engineers, scientists and astronauts to get another small concession to 'public affairs photography' for 'my friends in the press'. . . . Part of the 'right stuff' image was not to care about that sort of thing and the only argument which worked was to equate it with 'continued Congressional support'

. . . those kind of touristy snap shots were not programmed into the time line. They were supposed to be taken as a matter of course in among the 'scientific' or 'documentary' photography.

But, he added, 'they almost always were.'[35]

Each flight from Gemini 9 onwards had a photographic plan. For the Apollo missions this became a formal 'Photographic and TV operations plan' which set out in detail the main photographic requirements and added a list of possible 'targets of opportunity', such as pictures of the Earth and the astronauts themselves. Cameras were adapted for space work, with big film spools that avoided the need to change films and long levers that could be worked wearing gloves. Viewfinders were done away with altogether, as Underwood explained: 'you couldn't distort your body very well to look out the window of the spacecraft through a viewfinder, so they were taught to shoot from the hip.' The astronauts practised using cameras on their frequent flights in T-38 training aircraft, and on mission simulators. The stripped-down Hasselblad became the standard manual camera of the first space age. 'We brought the Hasselblad folks in to tell them what was a good picture and what was a bad picture,' recalled Underwood. 'That became integrated in their minds when they were in space.'[36] The choice to use hand-held cameras, rather than to adapt fixed surveillance cameras such as those used on U-2 spyplanes, had a deeper rationale behind it: 'we didn't want to get into the security business. Everything NASA did was in the public domain . . . . We wanted pictures that recorded it the way the astronauts saw it.'

Earth scientists who had seen orbital pictures would contact Underwood with specific requests for others. He would work out when the spacecraft was passing overhead in daylight and alert the astronauts. The detailed plan for photography was hidden in the spacecraft; only one astronaut knew where it was. 'I called it "photography for insomniacs",' said Underwood. 'In those days, astronauts didn't sleep, because they're looking at this Earth out there. No-one had ever seen this magnificent thing before. . . . The flight controllers couldn't understand why these pictures came back during sleep cycles.'[37] Sitting at the unused weather office console during missions, Underwood had access to weather satellites and other information and was able to remind the astronauts when a photographic target was coming up: 'It was quite informal. We wanted it that way and the crews wanted it that way.'

An editor of *National Geographic*, John Schneeberger, remembered Underwood from this time. 'He would bawl all those astronauts out about not doing well enough up there. He was a gutsy guy. His persistence and tenacity can't be overlooked.' 'After a while when I realized the value of the pictures,' said Underwood, 'I would tell some of them, "You know, when you get back, you're going to be a national hero." ' This would only last a short time, and the data they brought back would soon be obsolete too. ' "But these photographs, if you get great photos, they'll live forever. Your key to immortality is in the quality of the photographs and nothing else." Some of the guys would say, "Oh, Dick, you're crazy." And then the next day they'd say, "You know, you're right. I'll get you great pictures." '[38] Earth photography inserted a human touch into the corporate culture and technological routines of NASA; it gave the astronauts, at last, a little bit of freedom.

By robot eyes

The first photograph of the whole Earth was taken not by astronauts but by an ingenious orbiting photographic laboratory called Lunar Orbiter. While the Gemini astronauts were practising the space maneouvres that would be needed to get men to the Moon, robot pathfinder probes were already visiting the Moon to look for possible landing sites. Lunar Orbiter was the most sophisticated of these. It was operated not from Houston but by NASA's site in Langley, Virginia. Its job required high-resolution photographs, not wobbly television pictures, but unlike manned missions it could not be brought back. So how to get photographs back from the Moon in an age before digital photography?

The solution was ingenious: take black-and-white photographs, develop them on board, scan them with an electronic dot that could register differences of light and shade, transmit the impulses back home, and reassemble the picture on Earth. To avoid liquid chemicals sloshing about inside the probe, a dry development process was devised: the film strip was brought into contact with a roll of sticky

'Bimat' developer and then cured by a little electrical heater. The idea had originally been developed for military reconnaissance satellites on the secret Corona programme, where it was regarded as too complex to be relied upon. A back-up system was devised that involved releasing the negatives from the satellite in a re-entry vehicle and intercepting them with a C-130 transport aircraft.[39] No such back-up was possible from the Moon. To take a photograph, the controllers on Earth had to orientate the spacecraft correctly and then send a signal to activate the camera. Each picture was processed as it was taken. When the photography had been completed the film would be scanned and the signal transmitted, line by line, a quarter of a million miles, over a connection considerably slower than the slowest dial-up internet: two black-and-white pictures took forty-five minutes. Since the probe went behind the Moon every two or three hours, all this required careful timing, not to mention the full resources of NASA's global telemetry system.

Lunar Orbiter could only take 211 pictures, each of which was precious to those planning the manned Moon landings. Holiday snaps of the Earth were not at first on the agenda. There was no mention of photographing the Earth at an advance symposium, or even in the pre-mission press briefing. The final allocation of photo-frames was made at a key planning meeting in June; not one was given to the Earth.[40] Behind the scenes, however, NASA staff seem to have been dwelling on the idea of photographing the Earth. One possible source of inspiration was a campaign waged at college campuses over the spring by the future creator of the *Whole Earth Catalog*, Stewart Brand, asking: 'Why haven't we seen a picture of the whole Earth yet?' The campaign was briefly reported in the California press, and a number of button badges bearing the slogan reached NASA personnel, but not necessarily at Langley; whatever impact the campaign may have made at the time, no trace of it seems to survive in NASA's archives. The recollection of Richard Underwood is that 'the desire to have Lunar Orbiter 1 get a photo of Earth was a mutual one by all concerned at NASA. That included the Headquarters guys in Washington as well as us here at the Manned Spacecraft Center, and the Boeing guys, too. It was a "no brainer" as far as we were all concerned.' Langley, however, was in operational control.[41]

To tinker with a probe orbiting the Moon was a risky business. Lunar Orbiter was set up to take photographs straight downwards from a fixed camera. The entire probe would have to be turned around by a signal from the Earth and ordered to take a photograph looking over its shoulder at the marble-sized Earth. It would promptly disappear behind the Moon before anyone knew whether it had worked, and would have to be repositioned later. All this would happen early on in the mission, before the photographs of landing sites had been secured.

There was another problem: NASA managed its contractors through incentive schemes. Boeing was paid to do exactly what it said on the contract, and the contract didn't mention anything about photographing the Earth. NASA asked the question, but Boeing's programme manager Robert J. Helberg refused to take the risk. There was a chink of light, however. Lunar Orbiter had been designed to be 'tweaked' as it went along, to respond to what they found on the Moon. Boeing's incentive scheme had been deliberately kept open-ended: 'we gave them a bonus for useful, photographic data that did not attempt to define useful and said it would be determined unilaterally by the NASA evaluation board,' explained Lee Scherer, NASA's programme manager. As Lunar Orbiter was its first major project for NASA, Boeing in the end proved willing to please. According to the official history,

> The understandably conservative Boeing stance was changed through a series of meetings between top NASA program officials, including Dr. Floyd L. Thompson, Clifford H. Nelson, and Lee R. Scherer. They convinced Helberg that the picture was worth the risk and that NASA would make compensation in the event of an unexpected mishap with the spacecraft.[42]

The final decision was taken once Lunar Orbiter had reached the Moon. It had to take a photograph every so often to keep the film moving freely through the system, so a number of 'targets of opportunity' had been written in to use up the spare clicks. These had not originally included the Earth, but the developer was drying and getting sticky a bit quicker than expected. Could not the Earth become a 'target of opportunity'? Scherer recalled:

This is our first mission. We are all out at JPL [NASA's Jet Propulsion Laboratory in California] and we had just gotten our early pictures of the Apollo sites back and somebody came running up and said, 'We just made a calculation and as we drop behind the Moon on the next orbit . . . we can take a picture with the Earth to the Moon [sic] and this would be an extremely [interesting] picture.' And Boeing says . . . 'This is a tremendous risk. We might lose the whole mission and this tremendous incentive we have. . . . Why should we do this?' So we sat down for an hour around the table and . . . the vice-president of Boeing got up and says, 'To hell with it. It is a public service. It might be tremendous.'

We took it. It got tremendous play. Mr Karth [Joseph Karth, chairman of the congressional committee on space sciences] called me up there, in the mission, and says, 'What's this about you taking a picture of the Earth?' And I gave him a whole bunch of reasons why we did it, and – very defensively – he says, 'well, I don't give a damn why you did it, but me and 200 million other Americans thank you.'[43]

So it was that at 16:35 GMT on 23 August 1966, Lunar Orbiter took the first view of the Earth from the Moon. Langley issued a press release, explaining that

the purpose of the photograph was to obtain data, long of interest to scientists, on the appearance of the Earth's terminator (line dividing sunlit and shadowed portions of the planet) as viewed from a distance of about 240,000 miles. While from Earth the Moon's terminator is a sharply-defined line, atmospheric effects are expected to diffuse the sunlight and yield a view of the Earth's terminator as a gradual shading from light to dark.

A second, lower resolution photograph was taken two days later. The scanned images were downloaded to the Deep Space Network tracking stations at Goldstone, California, and Robledo De Chavela, near Madrid. Even then the process was not complete. The signal had to be copied onto videotape and projected onto a kinescope, where it was filmed with a movie camera. The film negative was then sent to the Eastman Kodak processing lab at Rochester, New York, cut into strips and processed into individual prints, which were in turn sent to Langley for analysis and publication. It was a while before the pictures were ready. The first, high-resolution photo was released to the press on 10 September as 'the world's first view of the Earth taken by a spacecraft from the vicinity of the Moon'.[44]

A photograph of the Earth from the Moon presented a problem of perspective. There is no 'up' and 'down' in space, so which way were the photographs of Earth to be printed? Lunar Orbiter, like all NASA spacecraft, left the Earth in an equatorial orbit (east to west) and orbited the Moon in the same plane. Looking back, Earth's North Pole was at the top and the sunset line – the 'terminator' – ran from top to bottom. This, naturally, is how all the photos of the Earth alone have been printed, with north at the top. The Earth as seen from the Moon, with the lunar landscape in views, poses a problem however. If we keep the North Pole at the top, then the Earth appears at the side of the Moon.

This was how the photograph was described in the caption issued with it: 'The Earth is shown on the left of the photo with the U.S. east coast in the upper left, southern Europe toward the dark or night side of Earth, and Antarctica at the bottom of the Earth crescent. The surface of the Moon is shown on the right side of the photo.' NASA wanted people to understand that the Earth looked different from space. In the same way, the official history of the Apollo programme printed it vertically and described it as 'a disarming view of the Earth'.[45] When the photograph was reissued to the press in March 1967, viewers were instructed to hold it with the Earth on the left and the Moon on the right side. Despite the care which was taken in the supporting text to explain that the Earth should be on the left of the Moon, on the front of the copies of the photograph released to the press the caption was printed below in the usual way, in landscape rather than portrait.[46] This was how the press chose to print the photo: in landscape format with the Moon below and the Earth above it, like a reversed image of the Moon above the horizon on Earth.

When the picture was re-released in March 1967 (perhaps to provide a ray of hope in the gloom following the Apollo 1 fire), a prophetic sentence was added: 'This is the view the astronauts will have when they come around the backside of the Moon and face the Earth.'[47] The wording implied, optimistically, that the Earth was rising. In fact Lunar Orbiter had taken the photo just before disappearing behind the Moon: this was not yet Earthrise but Earthset.

Even the 'right' way up, however, the Earth as seen from Lunar Orbiter just didn't look like home, in the way that it later would from Apollo 8. Senator Joseph Karth, who had been so excited over the picture, had earlier that year been talking about the 'need to view the world environment as a whole', but among many speeches in this period on both the space programme and its value for Earth science there is no mention of the impact of the Lunar Orbiter photograph.[48] In black and white, seen against the Moon, the Earth really was just another planet. Without colour, without a human eye behind the camera, without an astronaut to describe the setting, it came over as just a page from an astronomy textbook. The more tightly curved lunar horizon tilted to one side, the desolate sharpness of the lunar landscape, and the large size of the Earth relative to the Moon, all reinforced the sense of other-worldliness.

NASA's popular publication *Exploring Space with a Camera* encouraged readers to see the Earth in the context of the solar system and of space travel, as a precursor of future progress: 'by this reversal of viewpoint, we here on the Earth have been provided a sobering glimpse of the spectacle of our own planet as it will be seen by a few of our generation in their pursuit of the manned exploration of space. We have achieved the ability to contemplate ourselves from afar.' Beaumont Newhall, in his 1969 history of aerial photography, *Airborne Camera*, thought the Lunar Orbiter photograph was a sensation because 'for the first time we no longer looked *down* upon the Earth, but *at* the Earth'. Arthur C. Clarke considered that 'for many millions of terrestrials, their first glimpse of this photograph must have been the moment when Earth really did become a planet'.[49] This was not about finding the home planet, but leaving. It was not just a view of Earth from space; it was a view of Earth from the future.

At first NASA didn't seem to grasp the significance of what it had. NASA's scientific thinking about the whole Earth was fixed on what it called 'terminator studies': the study of atmospheric phenomena and surface detail in depth at the point where morning and evening shadows threw it into relief. For most people, however, it was hardly news that the sunset line was expected to show 'a gradual shading from light to dark'. The letters of congratulation sent out to his staff by the chief of the Lunar Orbiter programme, Oran W. Nicks, didn't mention the first photo of Earth among the list of technological achievements. The science journals which published summaries of the Lunar Orbiter findings ignored the Earth photographs almost completely. The various NASA publications on Lunar Orbiter which followed – there were five Lunar Orbiter missions in all – generally treated the Earth photos as a secondary objective, or at most 'a spectacular Lunar Orbiter by-product'. The press release issued to mark the end of the mission series failed to mention the Earth photographs among the achievements, and the official volumes of photographs which followed were mostly catalogues of craters. Only in materials prepared by the Public Affairs Office did NASA lead with the Earth photograph.[50]

On the next mission, Lunar Orbiter 2 took a magnificent panorama of the crater Copernicus. For astrofuturists, it was striking because of its resemblance to a widely reproduced painting by the 1950s' space artist Chesley Bonestell. More subtly, the name 'Copernicus' implied a historic perspective on the universe in which the Earth was de-centred – or, in this picture, absent entirely. It was this, rather than the photo of the Earth from the Moon, that was billed as 'the picture of the century'.[51]

In October 1966, a few weeks after the Earth from Moon photo was released, a NASA engineer posed a question: could one of the next Lunar Orbiters obtain photographs of the Earth on its way to the Moon? Leon Kosofsky, the chief programme engineer, took a couple of weeks to consider, but came down against. The probe could not be switched on until it was safely clear of the Earth's radiation belts, by which time it would be 60,000 miles (100,000 kilometres) away and the resolution would be no better than two kilometres: NASA was still thinking about surface detail, rather than the whole Earth. In any case the exposure was set for the Moon, so photographs of the sunlit Earth would be badly overexposed. But the main problem was the photographic system. Once it started taking photographs, it had to take one every eight hours to keep the sticky developer strip moving cleanly through the mechanism; in the ten days before it could photograph the first lunar site, thirty exposures would be wasted. The technical limitations that had made it possible to get a free photo of the Earth from lunar orbit made it too costly to take one on the way out. If photographs of Earth were deemed important, the mission objectives would have to be changed.[52] They were not; the Moon came first.

Between them the five Lunar Orbiters managed to snap the Earth in other perspectives. Lunar Orbiter 4, nearly 4,000 miles beyond the Moon, took a striking photo of the crescent Moon and crescent Earth side by side; to those on the Earth, by contrast, the Moon was nearly full. It prefigured the more famous Earth and Moon photo taken by the Voyager probe in 1977. On 8 August 1967, the very last Lunar Orbiter took a picture of the Earth – just the Earth. Once again it was an afterthought, using up a test frame. The black-and-white picture showed a fuzzy blob of light, but it was recognisably the Earth, taken alone and for its own sake. Finally, 'on one of the last missions,' recalled an official, 'we took the orbiter and cocked it so the Sun glinted off the solar panels that shine on the Earth just to see if anybody could see it'. They could, and the photograph shows a tiny glittering gadget a quarter of a million miles away.[53]

Unexpectedly, Lunar Orbiter's Earth photographs pushed the main programme forwards. The exercise showed that the probe could take oblique photographs across the lunar surface, as well as straight down – that is, views as well as mapping shots. Oblique photographs were programmed into future missions because of the detail and contours which they revealed. Boeing, which made the probes, finally received a 75 per cent bonus (a healthy $2 million) for achievements, including the 'one historic photograph' of the Earth.[54] So good were Lunar Orbiter's photographs of the Moon that they were still useful forty years later. On 29 October 1966, its mission accomplished, Lunar Orbiter 1 was nudged out of orbit by a signal from Earth to crash onto the far side of the Moon. Perhaps one day some lunar archaeologist will retrieve the remains of the first camera to photograph the Earth.

When Lunar Orbiter first arrived in lunar orbit, another probe called Surveyor was already sitting on the lunar surface taking photos. It had landed two months earlier, in June, and it wasn't interested in the Earth either. The early Surveyors took only black-

and-white pictures as engineers thought these would yield better data – 'an idiotic decision by some guys with three Ph.D.s apiece', thought the NASA photographer Richard Underwood, who lost the argument over colour. But for the third Surveyor there was a compromise: three colour filters would be taken to produce three separate images which, combined back on Earth, would look like full colour. In April 1967 Surveyor 3 managed to take a picture of the Earth eclipsing the Sun. The Earth is about four times bigger than the Sun when seen from the Moon, so it was a long eclipse and there was time to adjust the mirrors while the Sun was safely obscured. The mirrors had to be swivelled to their limits, right down on the horizon, but the picture just managed to get the Earth in the frame. The darkened disc appears dotted around with beads of refracted light.[55] A few days later it took an 'after' picture, showing Earth as a crescent, the first image of Earth in colour; in low resolution, however, the Earth was barely recognisable and it made little impact.

In January 1968 Surveyor 7 landed on the lunar highlands near the great crater Tycho, equipped this time with a colour TV camera. The camera itself was fixed but it was surrounded by a set of seven swivelling mirrors. During its first long lunar day (9–22 January 1968) Surveyor took a series of ten photos of the crescent Earth high in the sky, one for each Earth day, waxing and waning like the Moon.[56] Surveyor followed this up with a twenty-four hour series of photographs of the Earth rotating. It was described by NASA as 'a series of meteorological photos'. Neither feat seems to have generated very much press coverage. Attention was focused on two tiny dots on the dark side of a fuzzy close-up of the half-lit Earth: laser beams aimed at the Moon from Kitt Peak, Arizona and Table Mountain, California.

One of Surveyor's scientific experiments turned out to have great significance for the future view of the Earth. When the polarisation of light was measured, the Moon turned out to be normal, but the Earth 15–20 per cent brighter than expected, especially over the oceans.[57] When the Earth later appeared a glowing blue to the Apollo astronauts, the scientists already knew why.

Unlike the crashed Lunar Orbiter, Surveyor 3 did find its lunar archaeologist. In November 1969 the second Moon landing, Apollo 12, touched down less than 200 yards away from it. The astronauts walked over to pay it a visit and brought some pieces of it back to Earth with them. The expedition wasn't filmed because one of the astronauts, less deft than the little Surveyor, had accidentally burnt out Apollo's TV camera by pointing it at the Sun. As Surveyor underwent decontamination back on Earth, on one piece bacteria were found. Although they may have been picked up during decontamination, it is thought that they could have survived their three-year journey away from Earth – a discovery which, if true, has important implications for the origins of life.[58]

A robot probe, then, was responsible for the first view of the Earth taken from another world. For NASA, as for most space-minded observers, what was important was that it was a picture of the Earth from the Moon, not that it was a picture of the Earth, full stop. It was the view of the imaginary space voyager, not yet of the homesick astronaut. NASA was still thinking about the Earth in terms of detail – weather patterns, surface resolution and 'terminator studies' – but the sort of equipment that could be carried on small lunar probes wasn't capable of resolving the Earth into anything more than an interesting astronomical detail. While Stewart Brand was asking to see 'the whole Earth', the phrase does not yet seem to have been heard at NASA. Lunar Orbiter had seen a phase of the Earth rather than the entire disc, and not something that was the Earth in any immediately recognisable way. We had still not seen a picture of the whole Earth – yet.

CHAPTER 5

Blue marble

Not until they were leaving home for the last time did the Apollo astronauts finally get a good view of the full Earth. The famous Apollo 17 'Blue marble' photograph was the result, and it has since become the most widely reproduced image in history. The previous Apollo missions had taken pictures of the Earth in part shadow, but not the full Earth. Colour photographs of the full Earth in fact predated the Apollo programme. Just as the Earth rose four times before the Apollo 8 astronauts saw it, so too several photos of the full Earth appeared before humanity really took notice.

Earth in colour

The problem with getting a good picture of the whole Earth was finding a place to take it from. The early weather satellites were not set up for it. They needed to see weather systems in detail, and to pass over several times a day, which meant orbits too low to take in the whole Earth at once. Mosaics of the whole disc had an artificial, two-dimensional feel. The communications satellites of the mid-1960s, such as Early Bird, had more potential. They had to remain stationary over one point on the Earth, in a geosynchronous orbit some 22,000 miles up, and they could easily see the whole Earth. But extra weight was astronomically expensive, and no one was going to spend money sending up a communications satellite with a camera hanging round its neck just to get a few pictures of home. Would-be photographers of the whole Earth would just have to borrow a camera from some other project.

The first recognisable colour picture of the whole Earth was a by-product of a military satellite known as DODGE (Department of

Defense Gravitational Experiment). This carried a camera designed by Clyde Holliday, the pioneer of rocket photography in the 1940s. It was testing out a new way of stabilising satellites. Normally they stabilised by spinning, but photography (not to mention weapons targeting) would be easier if they could be kept still. DODGE was equipped with three pairs of retractable metal booms with weights on the end, which were adjusted by remote control to balance the satellite in three dimensions. The long vertical booms spanned a hundred yards (90 metres), enough for Earth's gravity to exert slightly more pull on the bottom weight than on the top one: one third of a millionth of a pound more. This was just enough to keep the satellite pointing vertically downwards – the same principle that keeps one side of the Moon always facing the Earth.

At 21,000 miles DODGE orbited slightly below geosynchronous height, circling the Earth every eleven days. In order to work out exactly how it was orientated in relation to the Earth, DODGE needed a camera. This was a black-and-white TV camera set up to take a one-second exposure, which it then took six minutes to scan and transmit back to the ground. Holliday and the other technicians watched the first image of the full Earth build up, line by line, on a monitor back at the Applied Physics Laboratory at Johns Hopkins University in Maryland. In those pre-digital days the TV picture was then photographed by an ordinary film camera, printed and developed. Three coloured glass lenses, red, blue and green, were placed in turn in front of the camera on a command from Earth, and the three images taken at six-minute intervals were reconstituted into a single colour picture. A small colour test disc was fixed within the camera's field of view so that the colour could be checked for accuracy against an identical disc in the laboratory. Between shots the Earth would rotate slightly and the satellite would wobble, but with careful processing the results were impressive: the whole Earth, in colour for the first time.

Perhaps the most remarkable thing about this achievement was the silence that surrounded it. Unlike NASA, the Department of Defense did not usually publicise its projects. Although APL sent out a press pack, it was dominated by technical information about 'gravity gradient stabilization' and 'three-axis vector magnetometers'. Among the list of DODGE's 'secondary objectives' was 'to take colour TV pictures', but nowhere was it mentioned that these would be of the whole Earth. When the pictures arrived the *Washington Post* went into colour to celebrate the achievement, but there seems to have been little coverage elsewhere. Very few newspapers could print in colour, so for most readers the picture would have appeared little different from a weather photograph; in newsprint, Lunar Orbiter's picture of the Earth over the Moon was much more striking. *National Geographic* magazine helped Holliday process the picture and ran an article on the project. It quoted George Meredith's 1883 poem 'Lucifer in starlight': 'Above the rolling ball in cloud part screen'd . . .' The Navy's Director of Astronautics in charge of the project, Captain L. P. Pressler, offered an interesting prediction: 'This will mean a lot to civilians too. For example, with a cheap five-foot aluminium-dish antenna, any school could pick up educational TV from special satellites.' Satellite TV was still twenty years in the future, but it is worth placing on the historical record the fact that the very first satellite colour TV picture to be transmitted was of the whole Earth.[1]

NASA meanwhile was working on a high orbital satellite carrying a camera with four times the resolution of DODGE's: the clumsily named Applications Technology Satellite. ATS had been announced in the usual NASA way in 1965 as a purely technological project: a multi-purpose experimental satellite in a parking orbit 22,300 miles above a point on the equator. It invited bids for experiments to be carried on board. Dr Verner Suomi of the University of Wisconsin proposed a new kind of camera, the ingenious 'spin scan cloud camera'. This was a way of taking photographs through a spinning satellite. Running at 100 revs a minute for 20 minutes, the camera would take a 2,000-line picture of the Earth, each line covering a two-mile band of the surface, which would then be transmitted back to Earth to be reassembled electronically. The technique was so precise that it had to allow for the minute effects of nutation – the slight wobble of the Earth's axis over an eighteen-year cycle. A tube of mercury would absorb this 'nuta-tional energy' and so avoid blurring during the twenty minutes of exposure time.[2]

With support from other meteorologists, Suomi's proposal was accepted.[3] The US Air Force expressed an unsubtle interest in using ATS 'for photographing meteorological coverage of Vietnam', but was disappointed to be told that the satellite would be in the wrong place – maybe next time. When the full programme was announced the next year, the Goddard Spaceflight Center proclaimed that its 'spectacular ... horizon-to-horizon pictures of the Earth' would score 'a prestige gain for the United States in addition to achieving technological benefits'. ATS-I was launched from Cape Kennedy on 6 December 1966 and soon began sending back pictures, steady and consistent views of weather patterns across a third of a billion square miles of the western hemisphere, from Cape Horn to Hudson Bay. They showed the Earth in its own right, full and bright and with detail to which people could relate. 'A showstopper,' exclaimed the Washington *Evening Star*, impressed that it even showed a snowstorm over Washington.[4] 'A roaring success, with performance beyond my wildest dreams,' said Dr Suomi.

These first detailed whole Earth photographs were in black and white (thirteen shades of it), but Dr Suomi had already made a prototype colour version of the spin scan cloud camera. This was sent up on the third ATS satellite on 3 November 1967 into an orbit above the mouth of the Amazon. On 17 November NASA released an 'excellent quality' image of the whole Earth in colour. In the press, however, during the two-year gap between manned spaceflights, the spaceman's view of Earth seemed to have faded in favour of the weatherman's. ATS was celebrated for its 'dramatic display of changing weather patterns over the Earth' and 'high quality pictures of Earth's cloud cover' more than for its first colour picture of the whole Earth. Meteorologists were excited by the prospect of using colour to analyse weather systems in depth, watching the birth of twisters and seeing the Gulf Stream at work.[5] ATS-I remained in operation for six years and ATS-III for eight, providing weather pictures used in nightly bulletins across the United States, as well as relaying global television pictures of events such as the 1968 Mexico Olympics. Dr Verner Suomi went on to become chief scientist of the US Weather Bureau. On his death in 1995 photographing the Earth was regarded as his highest achievement.[6]

NASA itself at the time did present the ATS pictures as a view from space rather than just a planet-wide weather map. Its press release for the ATS-III picture read: 'the photo shows the entire disk of the Earth, a cloud-covered globe in the blackness of space'. Later the same month, NASA put together a series of black-and-white ATS-III photographs to show the Earth waxing and waning from crescent to full and back again, demonstrating that it really was a planet in space. The *Washington Post* was also in a thoughtful mood. Printing a 'planet portrait' from ATS-I on an inside page, it asked: 'What is the planet and could it sustain life? Does it have vegetation? It must have atmosphere, as indicated by the massive clouds. The planet is Earth.'[7] Already, as the first electronic views of the globe appeared, minds were beginning to think in terms of the whole Earth.

In November 1967, at almost the same time as ATS-III was transmitting colour scans of the globe, NASA brought back a real photograph of the whole Earth that was less widely noticed but in many ways more impressive. Apollo 4 was 'the first of the big shots', the test launch of the colossal three-stage Saturn V rocket that would take men to the Moon. Inside the capsule on top, looking out of the window, was not an astronaut but a 70mm Maurer camera. Every picture of the whole Earth so far had been electronically transmitted; this time, the aim was to take real photographs and bring the negatives back to Earth. Even three miles away the blast blew in the window of CBS television's mobile studio, subjecting the broadcaster Walter Cronkite and his technicians to a sound roar of 120 decibels as they struggled to hold everything in place. 'Go, baby, go!' Wernher von Braun was heard to shout.[8]

The rocket made two low orbits of the Earth before the third stage booster fired to send it out in a vast ellipse, peaking at over 11,000 miles (18,000 kilometres). The ascent took nearly six hours; the descent two-and-a-half, mostly in free fall. It re-entered the atmosphere at 25,000 miles an hour, heating outside to twice the temperature of volcanic lava. The charred capsule was winched out of the Pacific two hours after splashdown; safe inside were 700 photographs of the Earth. The prints were captivating. The Earth was a blue and white globe, more white than blue, its quarter crescent filling the frame, looking at once three-dimensional and ethereal, like a lamp reflected in a darkened window. The atmosphere appeared at the edge as a thin white eggshell, tapering to a feather point.

Beautiful though they were, Apollo 4's pictures didn't make much impact. NASA's first round of press information was all about the technical performance of the rocket; copies of the photographs were only available a little later. 'This is an actual view of the Earth as a human eye would have seen it,' explained the caption. Very few papers picked up the photo, not even the normally image-hungry *Life* magazine in its feature, 'Impact of the supershot'. Perhaps newspaper editors judged that it wouldn't look the same in black and white; perhaps *Life* needed a human story to go with it. The picture ended up on the cover of *The Last Whole Earth Catalog* in 1971. 'No one seemed to care about noticing it or publishing it,' recalled the editor, Stewart Brand. 'I think it was the shadow which frightened people. There are no shadows on our maps.'[9] The ATS image of the full Earth was more widely printed. Today, though, Apollo 4's ghostly image of the Earth's globe, pale and breathing, like a child in the womb awaiting its first human witness, has a peculiar fascination.

No Russian cosmonaut ever saw the whole Earth. This was not something that would have been predicted even as late as 1965 when Alexei Leonov became the first man to walk in space, but after that the Soviet manned space programme ran into trouble. But before America sent men to the Moon, Russia sent tortoises. The Russian equivalent of Apollo was called Zond, and it too had been carrying out unmanned tests. They had notched up three failures and one partial success when, in mid-September 1968, Zond 5 was sent looping around the Moon with a cargo of insects and tortoises on a trial for a manned lunar voyage similar to that of Apollo 8. On its way out it took a black-and-white photograph of the Earth, two-thirds full, from 56,000 miles out (90,000 kilometres). There was excitement when Britain's radio telescope at Jodrell Bank picked up voices from Zond 5, but these turned out to be cosmonauts on the ground practising communications via the spaceship. James Webb described it as 'the most important demonstration of total space capacity up to now by any nation'. Wernher von Braun talked of a 'photo finish' in the race for the Moon, while Jodrell Bank's director, Sir Bernard Lovell, thought the Russians would get a man there first. As for Zond 5, it splashed down in the Indian Ocean; the tortoises were recovered alive and well. It is possible that the first living creatures to visit the Moon are still alive somewhere in the former USSR.[10]

Zond 6 followed in November, anticipating Apollo 8 by making a superbly accurate voyage to within 1,500 miles of the Moon and landing on target in central Asia. Unfortunately, the cabin depressurised and the parachute failed; had there been any astronauts on board, they would have died. From the wrecked capsule, however, were recovered some damaged film negatives. Among them was the first Earthrise brought back on an actual photo, black and white and coarse-grained but beating that of Apollo 8. 'If a Soviet Zond crew had returned high-quality photographs of the Earth from the Moon three weeks before Apollo 8,' the astronaut Dave Scott has suggested, 'the balance of the race could very well have changed dramatically.'[11]

After the celebrations for the first Apollo Moon landing in July 1969, little attention was paid to the unmanned Zond 7 a month later. This was a pity as among its pictures of the Earth was a colour Earthrise which, unlike any American picture of the day, actually showed the full face of the Earth above the lunar surface, just touching the horizon. As there were only thirty-five photos in all, this must have been by judgement rather than luck; it was a remarkable achievement. There was also a photograph of the nearly full Earth alone. Centred on Arabia and showing much of Asia as well as north Africa, it anticipated the later Apollo 'Blue marble'. Sadly, however, Russian photographic techniques, and indeed budgets, were well behind those of the Americans; the Soviet spacecraft carried unsuitable and often out-of-date films. Nonetheless, these Russian photographs of the Earth deserve to be better known than they are.[12]

By human eyes

It was Apollo 8 in December 1968 that provided the breakthrough moment: Earthrise, seen for the first time by human eyes. It may have taken the crew by surprise at that moment, but thanks to the thorough preparation they had received from Underwood and his technicians they recovered rapidly enough to capture the Earth rising. Here, if anywhere in the manned space programme, was the justification for having a human eye behind the camera. The impact of the occasion lay not just in the image that was returned but in the experience which was communicated. Over the next four years, there were to be many such experiences.

The Apollo programme lasted from late 1968 to late 1972. During that time there were eleven Apollo missions: nine went to the Moon, and six landed on it. Twenty-four astronauts beheld the Earth from afar (three of them twice), the last being the crew of Apollo 17 in December 1972. Each of the Moon missions took its own photographs of the Earth, including some of the Earth rising beyond the Moon that were technically better than Apollo 8's hurried first shot. *The Times*, normally restrained in such matters, put its front page into colour for the view of Earth from the Moon taken by Apollo 10, which in May 1969 rehearsed every detail of the Apollo 11 mission except for the actual landing. Even so, when the crew of Apollo 11 saw the Earth rise over the Moon, reported Michael Collins, 'it was a truly dramatic moment that we all scrambled to record with our cameras'. In Richard Underwood's estimation, Apollo 11 in July brought back 'the best Earth rises and sets, by far', although all the interest was in the pictures of Neil Armstrong and Buzz Aldrin walking on the Moon.[13] On Apollo 11, however, the lack of official priority given to photography caused NASA's greatest embarrassment: the failure to get a picture of the first man on the Moon.

All of the well-known photographs taken to be of the first man on the Moon on Apollo 11 are of the second man, Buzz Aldrin, not of Neil Armstrong. This has been explained in terms of personality clashes between the two men, and perhaps arguments over who should be the first to step out, but the underlying cause was much simpler: NASA hadn't really thought about it. The mission planners had only agreed to take colour film when the Public Affairs Officer angrily asked how a black-and-white picture of the first man on the Moon would look on the front cover of *Life* magazine. The astronauts themselves understood the importance of what Aldrin called 'standard home pictures for the folks back on Earth', but the mission plan called for Aldrin to film Armstrong on the surface but not to photograph him: 'emphasis will be on photographic documentation of crew mobility, lunar surface features and lunar material sample collection . . . Armstrong and Aldrin will use the Hasselblad lunar surface camera extensively during their surface EVA [extra vehicular activity] to document each of their major tasks.'[14] But the Hasselblad was hooked on a bracket on Armstrong's chest and it wasn't easy to swap it over between men wearing bulky spacesuits in low gravity.

Armstrong photographed Aldrin with the flag and Aldrin was about to return the favour when President Nixon came on the line to congratulate the men and declare (rehashing MacLeish's words): 'for one priceless moment, in the whole history of mankind, all the people on this Earth are truly one.' So a president, thinking of Earthrise and keen to use the occasion to deliver a 'one world' style message, deprived the world of a photograph of the first man on the Moon. (In the astronauts' quarantine period before the launch of Apollo 11, recalled Michael Collins, Nixon had been refused permission to meet the crew 'because he might infect us with his germs'. After splashdown they were photographed on board ship talking with Nixon from inside a quarantine chamber: who, one wonders, was being quarantined from whom? Nixon would soon cancel the later Apollo missions.) 'We could also look around and see the Earth,' recalled Aldrin. It 'seemed small – a beckoning oasis shining far way in the sky'.

> We had hoped to take a picture which would include the Earth, the lunar module, and one of us. In order to do so, one of us would have had to lie down on his belly and get the proper angle, and getting back up would most likely have been quite an effort. But I did manage to take a picture of the lunar module and the Earth. It turned out to be a disappointment.[15]

After the return to Earth, once the photographs had been processed, the Public Affairs Office rang Aldrin up in the night to ask: 'Where's Neil?' Aldrin couldn't remember, and nobody during the mission had thought to check: 'My fault perhaps, but we had never simulated this in our training.' 'We simply did not spot the potential for missing good photography of Armstrong,' recalled the press officer Brian Duff. 'It was a series of simple human oversights.' Afterwards, the Public Affairs Office asked Richard Underwood, 'Why don't we say this picture by the flag is Armstrong? How do you know? You can't see his face or anything.' Some eight-year-old kid would be bound to spot it, advised Underwood. 'Nobody in the news media picked this up . . . we were told, "Don't mention it." '[16]

After Apollo 11 there was a long wait before the next televised Moon landing. Apollo 12 followed four months later, but when the TV camera failed early on during the first Moon walk, coverage abruptly ended. Apollo 13, famously, had a problem on the way out and never landed on the Moon. It became the most gripping suspense story of the decade as the crew (commanded by James Lovell of Apollo 8) struggled with a series of near-fatal technical failures before eventually returning to Earth with only just enough power and oxygen left. All the focus was on the astronauts, yet as they drifted almost helplessly round the Moon and back they took a haunting series of photographs of an Earth they might never touch again.

Not until January 1971, eighteen months after Apollo 11, was a Moon landing fully televised, to be followed over the next two years by three more. Underwood eventually persuaded the CIA to let NASA have one of the large aircraft reconnaissance cameras used by the CIA on the U-2 spy flights, together with the special developing equipment that could print out pictures six feet across. The camera was fixed to the outside of Apollo 15 and as they orbited the Moon the astronauts 'would occasionally tip it up and look at the lunar horizon', yielding one spectacularly good Earthrise. Alfred Worden had to do a spacewalk to retrieve the negatives on the way back. On Apollo 16 astronaut John Young took an unusual full Earth disc from the surface platform on the Moon, using ultraviolet rather than normal light. The photo was merely a test exposure for an astronomical camera, one of whose functions was to detect whether there was any trace of an atmosphere on the Moon (there was none). It captured, in lurid artificial colours, the Earth's geocorona; behind were stars like dust, normally drowned out by the blue glow of the Earth.[17]

Blue marble

The impulse for a good photograph of the full Earth from Apollo seems to have come from outside NASA. In January 1972 James Fletcher, NASA's chief, and other senior officers had lunch with the leading people at *National Geographic* magazine. According to George Low, 'the National Geographic people told us that it was their impression that we had not taken as many pictures of the Earth traveling to or from the Moon as we had on earlier flights. They felt that this kind of photography continues to have great value.' Low asked his team to 'make sure that we have an adequate number of good Earth pictures, particularly if the geometry is such that we can get some of the nearly "full" Earth.' His deputy confirmed *National Geographic*'s impression:

of 350 distant pictures of Earth during the Apollo programme, only seventy-eight had been taken after Apollo 11 and six on the most recent flight, Apollo 15. 'There has not been a scientific requirement for large numbers of these pictures,' he explained, 'and there has been little demand from the press for them, in that for their purposes, the photos are largely repetitive.' The full Earth was expected to be visible for ninety minutes early on in the next flight, Apollo 16, but this was a busy time for the crew and as the Earth was designated only as a low-priority 'target of opportunity' the picture was never taken.[18]

There was a practical problem with getting a photograph of the full Earth. Apollo had to land on the near side of the Moon at a time when the Sun was at the right angle: not so high as to bleach out the surface features, but not so low as to dazzle the astronauts. This meant that the Earth too would be partly in shadow as seen from the Moon. Apollo 17, however, was heading for a region of the Moon near the edge as seen from Earth, and unusually it also launched at night. It therefore blasted out of Earth orbit towards the Moon around the middle of the day from a point over Madagascar. As they left home, the crew had a superb view of the full disc of the Earth, lit from horizon to horizon. Behind the camera was Harrison (Jack) Schmitt, a geologist and a geophysicist who, according to Underwood, 'understood the essential value of pictures of the planet Earth as you moved away from it' and was 'the only one to do a decent job with that'. 'I kept telling Jack Schmitt, who was a geologist, "that will be the classic picture. Make sure you get it after you go translunar," and Jack worked it into his schedule . . . that one's at 28,000 miles. That's a perfect picture and he aimed it beautifully.'[19]

Schmitt, the author of this iconic photo, was himself the least iconic of the astronauts. As a civilian he didn't really fit into the test pilot culture of the other astronauts; 'there was unbelievable friction between him and the pilots,' recalled one colleague. The others called him 'Dr Rock', and made fun of his single-minded sample collecting. His commander, Gene Cernan, had already flown round the Moon on Apollo 10 and been deeply affected by seeing the Earth rise. On the surface, while Schmitt was chipping, writes Cernan,

the Earth kept drawing my gaze away from the bleak surface, and reality felt like an hallucination. I had already seen it many times, but was still mesmerized by the most spectacular sight of the whole journey ... I made one more attempt to get Dr. Rock to realize he was on another world ... 'You see one Earth, you've seen them all.' It was typical of his droll humour, but I was almost disgusted with the blasé reaction, because I felt any human being should have been awestruck by the sight.[20]

Perhaps Schmitt knew that he could afford to be blasé, however, for he already had in the bag the best possible view of the Earth, and his rock-chipping was helping to raise Earth awareness in another way. The theory making the running at the time of Apollo 17 was that the Moon was actually a chunk of the Earth that had been knocked off by a gigantic impact during Earth's formative period, a theory that has since been generally accepted. As he resisted Cernan's entreaties to gaze at the Earth, Schmitt knew something that Cernan did not: in a sense, they were already there.

For Richard Underwood, the 'Blue marble' was the picture he was most proud of: 'More people have seen that photo than any in the history of mankind, and I saw it first. I was the first person to see that photograph. It was wet in a processor in Building 8. When I saw it I said, "Boy, that's it." ' Once it left the processing lab and was released copyright-free through NASA's Public Affairs Office, the 'Blue marble' became the property of the whole world. NASA released it on Christmas Eve 1972 – four years to the day since Apollo 8's Earthrise. But whereas NASA had had an astrofuturist story easily to hand about man's first sight of the Earth from another planet, it didn't have much to say about a picture of the Earth that was just – well, home. The caption supplied by the Public Affairs Office noted that it was the first photograph in which no part of the Earth was in shadow, and pointed out the land masses in view. *Aviation Week and Space Technology* confined its observations to the 'amorphous nature of cloud patterns'. The official Apollo 17 mission report made no mention at all of the photograph, although it did find space to record details such as the problems of 'gastrointestinal gas' which afflicted the astronauts ('at no time were the symptoms severe enough to interfere with the opera-tional duties of the crew': the mind boggles).[21]

NASA, perhaps anxious to overcome the prevailing sense that the manned space programme was at an end, continued to direct attention away from the Earth and towards space. When the Apollo 17 astronauts made their now ritual appearance before Congress, they didn't mention the 'Blue marble' photograph. Harrison Schmitt, the photographer, explained that through technology man had now 'evolved into the universe'. Reminding the Congressmen that the symbol of Apollo 17 was the American eagle, he vowed: 'The responsibility which we have toward the freedom of mankind throughout the universe will never, never be compromised.' Some newspapers shared the astrofuturist perspective, but the Earth was always somewhere in the picture. Before Apollo 17 blasted off, the *Miami Herald* had written that 'men's perception of their place in the cosmos underwent an epochal change four years ago. The new view came from that picture of the Earth, wreathed in cloud cover, rising over the barren surface of the Moon.' For the *Kansas City Star*, the view of distant Earth validated the progressive view of the space age: 'Man . . . escaped the prison of his planet. And in looking back across the void, understood that it was a prison only if he let it be. That stunning perspective, even had there been no rocks to carry home, would have been worth the price of the trip.'[22]

Others were just glad to get back home. 'Thank God, the crazy Apollo business is over,' commented William Hines afterwards in the *Chicago Sun-Times*. On its cover, the *New Yorker* carried a drawing of the Earth and Moon together; while a lonely flag stood on the Moon, the Earth celebrated with a party hat and a squeaker, as if to say 'the real party's over here'. Although Apollo had solved no earthly problems, commented the *Wall Street Journal*, 'surely the recognition of the Earth's aloneness stressed so eloquently by the Moon exploration and the photographs from space has been in no small way responsible for the ecology movement.'[23] Whether they stressed the view or the journey, all of these comments had one thing in common: they were written before the 'Blue marble' was released, and they had in mind the view of both Earth and Moon together. Four years on, Earthrise had lost none of its power.

The solitary 'Blue marble' view from Apollo 17 seems to have taken longer to work its way into consciousness. Like Earthrise, it was released just after the editorialising was over, but by the flight of Apollo 17 the Earth from space was a familiar sight. The new image was the finest yet, but it was not news in the way that the first had been. Its special qualities took time to sink in.

Because of the logistics of lunar travel and satellite positioning, most familiar images of the Earth had been taken from the equator or further north. Most also had featured north America and the Atlantic. The 'Blue marble' picture decentred the traditional map of the Earth. The viewpoint was over Africa, an equatorial continent notoriously diminished in size by traditional map projections which stretched the polar regions in order to achieve navigational rather than visual accuracy. The Peters projection of the globe, which sought to equalise the land areas of the Earth and so combat the unthinking self-centredness of the developed world, was becoming widely known, but its distorted outlines made it seem unreal. 'Blue marble' offered equity without distortion. Now Africa appeared at the centre, with Europe and Asia visible as narrow bands towards the northern horizon. The winter tilt towards the southern hemisphere revealed the hidden continent of Antarctica, never before seen by astronauts. It also showed a high proportion of sea, as if to remind people that not only the Earth but even the habitable part of it was only a small part of a more significant whole. The whole was wreathed in white clouds, showing that it was a planet rather than a globe, but not so many clouds as to obscure the continents. Although no one found the words to say so at the time, the 'Blue marble' was a photographic manifesto for global justice.

Taken from relatively close to the Earth, 'Blue marble' also showed the land masses with greater clarity than most whole Earth photos. Viewed from greater distances, however, virtually all the familiar detail disappeared. The Earth as seen from the Moon was not a high-level weather map, as shown by the first geostationary satellites. Nor was it a geographical globe, as had been portrayed in NASA's advance profile of the Apollo 8 mission. Rather, it was an abstract composition in blue and white. Compared with what had been anticipated it was less like a photograph and more like an impressionist painting, with the familiar detail obscured and the deeper mysteries of nature displayed in a hypnotic blob of colour.

By the time of Apollo 17, the view was setting in that the first space age was over, and that its star was the Earth. The house magazine of the Aerospace Industries Association ended its retrospective review of the Apollo programme with the Apollo 8 Earthrise, and made what, coming from this source, was a remarkable comment: 'If those photos of lonely Earth wandering through the black of endless space planted a seed, that Earth is a tiny spaceship whose supplies are not replenishable and whose crew must work together for survival, that could be the greatest benefit.'[24] This might, thought the writer, take a thousand years or so to show up. In fact, it had already begun.

CHAPTER 6

An astronaut's view of Earth

'How beautiful our Earth is!' Yuri Gargarin exclaimed as he became the first man in space.[1] The well-known limitations of astronaut language have obscured the fact that many of the space travellers who viewed the Earth, particularly from the Moon, felt that they had undergone profound inner changes.[2] 'No one who went into space wasn't changed by the experience,' observed Charles Berry, the astronauts' chief doctor. 'I think some of them really don't see what happened to them.' Edgar Mitchell of Apollo 14 agreed: 'No man that I know of has gone into space . . . and has not been affected in some way.' Almost all of the astronauts interviewed in the 1980s by Frank White for his important and pioneering book *The Overview Effect* spoke of 'seeing the Earth from orbit and being emotionally unprepared for the experience'.[3] The travellers were impressed by the invisibility of human activity, particularly political boundaries, by the uniqueness and fragility of 'Spaceship Earth', and by the obvious common interests binding together those who lived upon it. Many felt an almost unbearable responsibility to communicate what they had learned to the rest of the human race, particularly its political leaders.

Considering that virtually all the Apollo astronauts came from similar white Anglo-Saxon male test-pilot backgrounds, and had shared years of intensive training, their own lives afterwards were remarkably varied. Some enjoyed successful public or business careers and long and stable marriages, all three Apollo 8 crew among them. Others variously experienced divorce, alcoholism, and religious or mystical transformations, or went into politics ranging from New Age Democrat to cactus conservative. Between them the small band of Apollo travellers founded three organisations aimed at spreading the awareness they had gained to the whole of humanity. Others wrote memoirs or poems, or painted pictures. As Russell Schweickart recalled, 'Many of us, on returning home from space, brought back the perspective of a lonely and beautiful planet crying out for a more responsible attitude from its most prolific partner. Strangely, we didn't talk about the stars much.'[4]

The twelve who actually walked on the Moon have gained most attention, but twelve more went there but did not land – six among the crews of Apollo 8, 10 and 13, and six more thoughtful command module pilots who waited alone in orbit for the Moonwalkers to return. Another five flew only on the Earth orbital missions, Apollo 7 and 9, enjoying similar views to those on the earlier Mercury and Gemini programmes but with more time and cabin space to appreciate them. Revisiting the astronauts' accounts of their own experiences, the biggest difference seems to be not between those who walked on the Moon and the rest, but between those who left Earth orbit and the rest.

Sitting in a tin can

Photographs of the Earth are powerful and fascinating, but a picture is not the full experience. Space shuttle crewman Don Lind explained: 'I have probably looked at as many pictures from space as anybody . . . so I knew exactly what I was going to see. There is no intellectual preparation I hadn't made. But there is no way that you can be prepared for the emotional impact.' The Russian cosmonaut Oleg Makarov put it like this: 'It's not just that the planet is piercingly beautiful when viewed at a distance; something about the unexpectedness of the sight, its incompatibility with anything we have experienced on Earth, elicits a deep emotional response.' Listening to recordings of Russian space missions, Makarov noticed: 'Within seconds of attaining Earth orbit, every cosmonaut, without exception, be they a dry, reserved flight engineer or a more emotional pilot, uttered the same sort of confused expression of delight and wonder.'[5]

The Earth was easy to see from orbit, and easier still when presented in a photo frame, but from much further away it was a different matter. For Michael Collins of Apollo 11, those who saw pictures of the Earth and then thought 'Oh, I've seen everything those astronauts have seen' were kidding themselves; an image alone was 'a pseudo-sight that denies the reality of the matter'. He described his own experience returning home with Apollo 11:

> I looked out of my window and tried to find Earth.
>
> The little planet is so small out there in the vastness that at first I couldn't even locate it. And when I did, a tingling of awe spread over me. There it was, shining like a jewel in a black sky. I looked at it in wonderment, suddenly aware of how its unique-ness is stamped in every atom of my body. . . . I looked away for a moment and, poof, it was gone. I couldn't find it again without searching closely.
>
> At that point I made my discovery. Suddenly I knew what a tiny, fragile thing Earth is.[6]

The cramped conditions and restricted views of the Apollo capsules made the Earth elusive. The cabin lights didn't help; Apollo 13, running on low emergency lighting, probably got the best view of Earth. On the journey out, recalled James Irwin of Apollo 15, the Earth shrank from the size of a basketball, to a baseball, a golf ball, and finally a marble, 'the most beautiful marble you could imagine'. For Bill Anders it was 'like a Christmas tree ornament'. When the crew of Apollo 10 took the first colour TV camera into lunar orbit, they at first had trouble finding the Earth in the capsule windows; 'ask the navigator' was Capcom's dry response. As Apollo 11 jour-neyed towards the Moon, slowly rotating, recalled Buzz Aldrin, 'the Sun, Moon, and Earth appeared in our windows one at a time'. He got the best view of Earth by floating a monocle in front of the window.[7]

On the surface of the Moon, James Irwin found that it was just as difficult to see the Earth from inside a bulky space suit as from inside an orbiting capsule. 'In the three days of exploration, there were a couple of times when I actually looked up to see the Earth – and it was a difficult maneuver in that bulky suit; you had to grab onto something to hold yourself steady and then lean back as far as you could.' 'To take a picture which would include the Earth, the Lunar Module, and one of us,' worked out Buzz Aldrin, 'one of us would have had to lie down on his belly.'[8] Gene Cernan's carefully framed photograph of Harrison Schmitt and the flag with the Earth in the background offers a rare impression of the Moonwalker's small world. From the Moon, the sight of the Earth could never be taken for granted; the astronauts had to learn how to look.

The smaller the Earth got, the more powerful it became. Michael Collins explained: 'There is definitely a completely different feeling. At 100 miles up you are just skimming the surface, and you don't get a feeling for the Earth as a whole. . . . You sort of have to see the "second planet" to appreciate the first.' For Gene Cernan, who went to the Moon twice, the Earth orbital and lunar missions were like 'two different space programs'. In orbit, you were still part of the Earth; from far away, it was 'like a multi-coloured, three-dimensional picture'. 'You look out of the window and you're looking across the blackness of space a quarter of a million miles away, looking at the most beautiful star in the heavens . . . and it's moving in a blackness that's almost beyond conception. . . . What are you looking at? What are you looking through? You can call it the universe, but it's the infinity of space and the infinity of time.' For James Irwin, the Earth was 'the only warm living object that we saw in space on our flight to the Moon'. Neil Armstrong, trying to sleep on his couch at Tranquility Base, was kept awake by the Earth shining down through his telescope 'like a big blue eyeball'; it was so small that he said he could blot it out with his thumb.[9]

The tension of space travel tended to heighten other emotions; in the case of Apollo 17, there was the added worry of a threat of launch pad attack by the terrorist organisation Black September. The long wait on the couch for take-off, followed by the gruelling physical experience of the launch and then the experience of floating silently in orbit only minutes later, marked a transition from normal earthly life. Once up, however, the astronauts could not relax. On Gemini 10, remembered Michael Collins, 'hardly an hour would pass without a fresh opportunity for disaster', but a more immediate fear was that of making a mistake in front of his colleagues: he 'felt this pressure, this awesome sense of responsibility weighing me down'. Walter Cunningham of Apollo 7 concurred: the astronaut's greatest fear was 'making an ass of yourself – especially in front of your peers'. Al Bean of Apollo 12 put it like this:

There's just the three of you in this little metal can, and you look out the window and see the Earth far away, and you realize that you can't intuitively get back home. . . . If something goes wrong, the alternatives go down to maybe two: either the ground is going to tell you how to get it home, or somehow you've got to figure it out with your computer.

For these overburdened astronauts, sheltering in routine, the emotional pull of looking out and seeing the home planet was perhaps all the greater.[10]

Russell Schweickart of Apollo 9 described the ascent to orbit:

As you pitch over, getting horizontal, you catch your first glimpse out the window of the Earth from space. And it's a beautiful sight. So you make some comment – everybody has to make a comment when he sees the Earth for the first time – and you make your comment and it's duly noted. And then it's to work, because you don't have time to lollygag and look out the window and sight-see.[11]

The work of the following days was punctuated by brief, frustrating sights of the Earth below. Gerald Carr of Skylab explained that 'you begin to get so involved with the details of what you're doing that I think you forget to look about you'. James Irwin of Apollo 15 wrote:

During this sort of flight, you are too busy to reflect on the splendour of space or on the secret awakenings that come from the inner flight that takes place at the same time. You have to try to register these experiences and examine them later . . . I had been so absorbed in preparing for the scientific flight that it never even occurred to me how high the spiritual flight could be.[12]

Astronauts drilled in technical procedures found it difficult to communicate their experiences and gained a reputation for dullness of language. They said things like 'main valve closed', and 'reading you loud and clear, Houston', as if they had been watching too many episodes of *Thunderbirds*. Even before Apollo 11 left for the Moon, a journalist speculated that it was 'headed for a rhetorical wreck'.

'We weren't trained to emote, we were trained to repress emotions,' explained a somewhat exasperated Michael Collins afterwards. 'If they wanted an emotional press conference, for Christ's sake, they should have put together an Apollo crew of a philosopher, a priest, and a poet – not three test pilots.' On Apollo 15, Dave Scott found himself 'unable even to begin to convey the wonder I felt looking back at the Earth from this distance . . . ' "Oh, this is really profound," I said. "I'll tell you, it's fantastic." ' He later learned that 'this diversion from the normally clipped and controlled transmissions expected of us slightly exasperated some at Mission Control'.[13] It was no wonder that astronauts tended to bottle up their experiences for later.

At times, the distant Earth made the gap between space and normal experiences seem unbridgeable. 'To stand on the lunar surface and look back at our Earth creates such a personal sense of awe that even Alan Shepard wept at the view,' wrote Gene Cernan. For himself, 'I was watching time go by on Earth, but time as we understand it did not really affect us on the Moon.' 'Where really am I in space and time?' he wondered. 'We were outside of ordinary reality,' explained James Irwin. 'I sensed the beginning of some sort of deep change taking place inside of me.' 'It is a pity that my eyes have seen more than my brain has been able to assimilate,' lamented Michael Collins. He remembered the incongruousness of listening from lunar orbit to President Nixon's rhetoric about Apollo 11 bringing 'peace and tranquillity to Earth' and then having NASA cut in to provide technical information. 'I never thought of this bringing peace and tranquillity to anyone . . . roll, pitch, and yaw; prayers, peace, and tranquillity. What will it be like if we really carry this off and return to Earth in one piece, with our boxes full of rocks and our heads full of new perspectives for the planet?'[14] Unprepared, preoccupied, confined and disorientated, some astronauts found that the sight of Earth hit with the force of a religious experience.

The eyes of the world

The most eloquent account of an astronaut overwhelmed by Earth is perhaps that of Russell Schweickart. Although he himself never went beyond Earth orbit, he had been deeply impressed by the way the Apollo 8 astronauts had read from Genesis 'in a sense to sacramentalize that experience and to transmit what they were experiencing to everyone back on Earth'. His flight, Apollo 9, was dubbed 'the connoisseur's mission', and Schweickart was in charge of the lunar module as it practised the difficult undocking and redocking manoeuvres that were essential for a lunar landing. It was an arduous mission which exhausted the crew; command module pilot Dave Scott had trained grimly for the possibility that he would have to return to Earth alone, leaving the other two marooned in the lunar module. Schweickart had to perform a long spacewalk. There were technical problems, and while the others sought advice from mission control, 'I had about five minutes just to look at the Earth and think about what I was doing, how I got there and what it meant.'[15] There was no window between him and the Earth; when he came to write about the experience years later he called the essay 'No frames, no boundaries'. He recalled Neil Armstrong's comment that from the Moon, the Earth was smaller than his thumb.

> A little later on the person next to you goes to the Moon. And he comes back and now he sees the Earth . . . as a small thing out there. And the contrast between that bright blue and white Christmas tree ornament and the black sky, that infinite universe, really comes through. It is so small and fragile and such a precious little spot in that universe that you can block it out with your thumb, and you realize that on that small spot, that little blue and white thing, is everything that means anything to you – all of history and music and poetry and art and death and birth and love, tears, joy, games, all of it on that little spot out there that you can cover with your thumb. And you realize from that perspective that you've changed.[16]

'During the second half of the mission,' recalled Scott, 'we had more time to reflect. Perhaps more than during the short and extremely successful mission with Neil on Gemini 8, I think I really appreciated during this second time in space the astonishing beauty of the stars and the planet Earth. At one point we dimmed the lights on the spacecraft so that we could get the best possible view.' They were rewarded with the sight of streaks of light in the atmosphere:

shooting stars, seen from above. They also talked a lot. The ex-test pilot Scott was a little surprised by Schweickart, whose background was as a research scientist at the Massachusetts Institute of Technology. 'He was a really cultured man. He had brought quotations from Elizabeth Barrett Browning and Thornton Wilder along on the flight', as well as a cassette of Vaughan Williams's cantata *Hodie* and Alan Hoyhaness's *Mysterious Mountain* which Scott (not liking classical music) hid in his suit pocket until near the end of the mission: 'He never forgave me for that.'[17]

Schweickart was not the only one to be struck by the absence of human boundaries on the Earth. Nearly all well-known atlases in this period were diagrammatic, structured by national boundaries; atlases using quality colour printing to show natural features as they might appear from space only began to appear at the end of the first space age. After over twenty years of human experience of space flight, the space shuttle astronaut Don Lind was still struck by how 'you can't see the boundaries over which we fight wars'. The 'feeling of brotherhood', he observed, remained one of the most common responses among his fellow astronauts. On another shuttle mission, Sultan Bin Salman al-Saud from Saudi Arabia, the first Islamic astronaut, reported: 'The first day or so we all pointed to our countries. The third or fourth day we were pointing to our continents. By the fifth day we were aware of only one Earth.' The cosmonaut Oleg Makarov wrote: 'Suddenly you get a feeling that you've never had before, that you're an inhabitant of Earth.' 'You don't look down at the world as an American but as a human being,' said Tom Stafford of Apollo 10. Alfred Worden made the same point during lonely orbits of the Moon on Apollo 15 by broadcasting the words 'Hello Earth, greetings from Endeavour' in many different languages. 'From out there,' said Frank Borman, 'it really is "one world".'[18]

Political leaders, thought Michael Collins, needed to see the unity of Earth for themselves: 'The pity of it is that so far the view from 100,000 miles has been the exclusive property of a handful of test pilots, rather than the world leaders who need this new perspective, or the poets who might communicate it to them.' When Senator Jake Garn of Utah became the first politician in orbit, on the space shuttle, he agreed: 'You certainly come to the recognition that there aren't any political boundaries out there. You see it as one world, and you recognize how insignificant the planet Earth is . . . if more people fly, there has to be more understanding of what I'm talking about.'[19]

For many astronauts the experience of looking down on creation deepened their faith in a creator God. James Irwin wrote: 'As we reached out in a physical way to the heavens, we were moved spiritually. As we flew into space we had a new sense of ourselves, of the Earth, and of the nearness of God.' Rare sightings of the Earth on his Moonwalks deepened the experience and gave him an 'overwhelming sense of the presence of God on the Moon'; when he encountered a technical problem in setting up an experiment, he prayed and immediately found an answer – 'God was telling me what to do.' 'Jim was deeply affected,' commented his fellow Moonwalker Dave Scott. 'Something real happened to him.'[20]

Gene Cernan of Apollo 17 also found renewed faith on the Moon. 'As I stood in sunshine on this barren world somewhere in the universe,' he wrote in his autobiography, 'looking up at the cobalt Earth immersed in infinite blackness, I knew science had met its match.' 'What I saw was almost too beautiful to grasp. There was too much logic, too much purpose – it was just too beautiful to have happened by accident. It doesn't matter how you choose to worship God . . . He has to exist to have created what I was privileged to see.' Few indeed were those who could say, with Cernan, 'I have seen the endlessness of space and time with my own eyes.'[21] The Saudi Prince Sultan Bin Salman al-Saud said: 'The minute I saw the view for the first time . . . I just said, in Arabic, "Oh, God", or something like "God is great", when I saw the view. It's beyond description. . . . It has changed my insight into life. I've got more appreciation of the world we live in.' Al Bean of Apollo 12, thinking about the famous 'Blue marble' photograph, said: 'I'm not a religious person . . . but I think, really, the whole Earth is the garden of Eden. We've been given paradise to live in. I think about that every day.'[22]

Perhaps the most thorough transformation was experienced by Edgar Mitchell, who flew to the Moon on Apollo 14. 'Something happened to me during the flight that I didn't even recognize at the time. I would say it was an altered state of consciousness, a peak experience if you will. I flipped out. . . . What is it that caused this? It was the view of the Earth.' Gazing on the distant Earth as the craft floated home, he 'suddenly experienced the universe as intelligent, loving, harmonious'. The idea of NASA's highly trained astronauts flipping over into some alternative kind of awareness has proved fascinating. The euphoric astronauts' narratives used in Al Reinert's 1989 film *For All Mankind*, backed with an electronic soundtrack by Brian Eno, struck some as having a decidedly psychedelic effect. Edgar Mitchell felt that the experience of seeing the Earth was so powerful that there was a 'danger of planetary drop-outs'. Had he but known it, a dazzling early NASA film of the Earth from space was already being marketed direct to the counterculture.[23]

Alfred Worden found that his experience of orbiting the Moon on Apollo 15 'changed his entire view of reality on Earth . . . gave him a profound feeling of rejuvenation . . . changed his life'. He wrote a number of poems about the flight, always returning to the impact of seeing the Earth. 'Now I know why I'm here: not for a close look at the Moon,/ but to look back/ at our home/ the Earth,' he wrote in 'Perspectives'. His poem 'Hello Earth' was printed in the NASA publication *Apollo Over the Moon* alongside photographs of the home planet.[24]

> Quietly, like a night bird, floating, soaring, wingless
> We glide from shore to shore, curving and falling
>   but not quite touching;
> Earth: a distant memory seen in an instant of repose.

From Spacemen to Earthmen

Those who travelled to the Moon often felt a responsibility to communicate what they had come to understand to the rest of humanity. 'I feel the responsibility of being a representative,' explained James Irwin of Apollo 15. 'Indirectly, everyone on Earth was a part of this flight.' Russell Schweickart, thinking about his spacewalk, said:

> You think about what you're experiencing and why. Do you deserve this, this fantastic experience? Are you separated out to be touched by God, to have some special experience that others cannot have? You know that the answer to that is no. . . . It comes through to you so powerfully, that you're the sensing element for

man. You look down and see the surface of the globe that you've lived on all this time, and you know all those people down there and they are like you, they are you – and you somehow represent them. You are up there as the sensing element, that point out on the end, and that's a humbling feeling.[25]

Perhaps an element of survivor's guilt was at work here. Schweickart had once had a brush with death in the company of fellow astronaut Ed White, who had described the end of his own spacewalk as 'the saddest day of my life'. In July 1966 Schweickart and White were taking off from El Paso, Texas in a high performance T-38 jet trainer when an engine failed; two tyres blew out, the front wheel came off and they overran the runway.[26] They were unhurt, but shaken: five months earlier two other members of the close-knit astronaut corps, Charles Bassett and Elliot See, had been killed in a crash on a similar jet. Six months later White himself died in the Apollo 1 fire; he never wrote up his own story.

Standing as the last man on the Moon, Gene Cernan felt: 'My destiny was to be not only an explorer, but a messenger from outer space, an apostle for the future.' He was frustrated: 'I was unable adequately to share what I felt. I wanted everyone on my home planet to experience this magnificent feeling of actually being on the Moon.' After realising how small and fragile the Earth was, declared Michael Collins, 'I determined in that moment that I would do all I could to let people know what a wonderful home we have – before it is too late. So I have a personal, simple message to pass on: There is only one Earth. It is a tiny, precious stone. Let us treasure it; there is not another one.' Al Bean left the astronaut corps to become an artist; many others could fly the space shuttle, but only he could paint what he saw on the Moon. On Skylab he had found it difficult to describe the sunrises in words ('sort of orangy-gold colour . . . a little bit blue'), but his pictures were deeply expressive of the experience of space: 'My dream now is to create a body of paintings that tell the story of Apollo,' he explained.[27]

Returning from his Moonwalk on Apollo 15, James Irwin was re-baptised at Nassau Baptist Church, Houston. He founded an evangelical organisation called High Flight, transforming himself (thought his colleague Dave Scott) from a good public speaker into a great one. 'Jim was deeply affected . . . something real happened to him,' observed Scott. High Flight was named after a poem of that name by John Magee, popular among pilots: 'And, while with silent, lifting mind I've trod/ The high untrespassed sanctity of space,/ Put out my hand, and touched the face of God.' Michael Collins took a copy of it to the Moon. Irwin's Apollo 15 crewmate Alfred Worden joined High Flight in 1975. Having discovered the famous 'Genesis Rock' on Apollo 15, which confirmed the common origin of the Earth and Moon more than four billion years ago, Irwin embarked on several missions to find Noah's Ark on the slopes of Mount Ararat.[28]

Edgar Mitchell went even further. On the Apollo 14 mission he had experienced what he called 'an explosion of awareness, an aha! a wow!' On the voyage home he had attempted experiments in extrasensory perception; back on Earth he reordered his ideas away from the 'scientific paradigm' in a more mystical direction. Mitchell left the space programme quite soon after returning to Earth and went on to found the Institute for Noetic Sciences at Palo Alto, California, a research, education and philanthropic organisation designed to promote the advance of human consciousness. Creative insights, he realised, happen 'when the brain is in resonance with the fundamental stuff of the universe'. He expected that the experiences of the astronauts would have much wider effects: 'getting outside of Earth and seeing it from a different perspective . . . will have a direct impact on philosophy and value systems'.[29]

The most common reaction among astronauts to the sight of Earth has probably been heightened concern for the environment. All those who have seen the Earth from orbit have been struck by how thin the atmosphere is; the apparently limitless blue sky turns out to be no thicker than the peel of an apple. Gerald Carr, who trained for Apollo but ended up commanding a Skylab mission, said: 'Most of the guys come back with an interest in ecology . . . you come back feeling a little more humanitarian.' Among those who have seen the Earth from the Moon the sense of its fragility seems to have been particularly acute. Michael Collins saw both.

The smoke from the Saar Valley may pollute half a dozen other countries, depending on the direction of the wind. We all *know* that, but it must be *seen* to make an indelible impression, to produce an emotional impact that makes one argue for long-term virtues at the expense of short-term gains. . . . Anyone who has viewed our planet from afar can only cry out in pain at the knowledge that the pristine blue and whiteness that he can still close his eyes and see is an illusion masking an ever more senseless ugliness below.

For some later space travellers the causation ran the other way; fifteen years before his space shuttle flight, Charles Walker had participated in the first Earth Day.[30]

Edgar Mitchell recalled sensing as he travelled back towards Earth on board Apollo 14 that 'beneath the blue and white atmosphere was a growing chaos . . . that population and conscienceless technology were growing rapidly way out of control'. His feeling was that 'the crew of spaceship Earth was in virtual mutiny to the order of the universe'.[31] Russell Schweickart went farthest in the environmental direction, joining the Californian counterculture. A year after Apollo 9, he contributed an unusual piece to *Bell Rendezvous* entitled 'Earth: Planet 3A of Sol', an assessment of the state of the Earth by an imaginary space traveller. He saw an Earth at 'a balance point in its evolution, teetering between potential greatness and colossal collapse', between 'warfare to protect and defend parochial boundaries' and 'intelligent control and management of the planetary resources'. He was most impressed by 'the younger Earth people of these advanced countries . . . angry, discouraged, loud, iconoclastic, revolutionary, and impractical . . . they are, in the last analysis, the only real hope for the planet'. As we will see in chapter 8, he went on to enter the counterculture in order (as he put it) to 'share an experience which man has now had'.[32]

It is possible too that some environmentally tinged comments made years afterwards owed something to the growth in environmental awareness after Apollo, which may have helped the astronauts articulate their original basic feeling that the Earth was fragile and precious. When he came to write his autobiography in 1988, Borman regarded Earthrise as 'the most beautiful, heart-catching sight of my life'. But at the post-recovery press conference he had said: 'I think one of the first comments was please pass me the camera. The events at that time were not nearly as dramatic as one might think.[33] Nearly thirty years later, James Lovell pondered the Genesis reading:

> Actually, looking back . . . brought [it] home more to me than . . . doing it at the time. We thought it was very appropriate at the time. We were so wrapped up in the Apollo that we missed a lot or didn't pay much attention to all the riots and the war that was going on, the assassinations. . . . Years later when we see that flight in the context of everything that was going on in 1968 in the United States, it had a greater impact on me.[34]

Ironically, among those least affected by going to the Moon was the man who took the 'Blue marble' picture: Harrison Schmitt, the geologist, 'Dr Rock' himself, who exasperated Gene Cernan by continuing to chip away at the Moon while the Earth hung hypnotically in the sky. The two men took turns to photograph each other against the American flag; Cernan took care to get the Earth in the frame, but Schmitt didn't bother. Addressing Congress after Apollo 17 returned to Earth he spoke not of the Earth but of technological evolution and the spread of American freedom into space. Writing later, he tried to face both ways at once: 'Like our childhood home, we really see the Earth only as we prepare to leave it. . . . The modern challenge, emphasized by our travels, is to both use and protect that home, together, as peoples of the Earth.' Schmitt was not alarmed by environmental change, commenting: 'I think the pictures make it look a lot more fragile than it is. The Earth is very resilient. Again, geologically, I see that. I know what blows it's taken. . . . And in fact those rapid changes on Earth may well have been what forced human evolution to where we are today.'[35]

Four years after his return to Earth, Schmitt became the Republican Senator for New Mexico, trouncing the incumbent Democrat in a surprise landslide. His politics were 'a maximum of individual opportunity with a minimum of government interference and a strong defense'. He was known as 'Moonrock' on Capitol Hill, with a reputation for being unworldly and lacking in social instincts and for looking on political issues with scientific disdain. Hostile to any form of energy regulation, he became one of the handful of 'cactus conservatives' who resolutely opposed even Ronald Reagan's compromises, describing himself as 'an ideology of one'.[36] But in a way Schmitt's apparent lack of transformation is perhaps not surprising; as a geologist, unlike his test pilot colleagues, he was already used to thinking in planetary terms.

Some other astronauts were little changed by their experience. Rakesh Sharma, the first Hindu astronaut (on the Russian Soyuz programme in 1984), said: 'My mental boundaries expanded when I viewed the Earth against a black and uninviting vacuum, yet my country's rich traditions had conditioned me to look beyond man-made boundaries and prejudices. One does not have to undertake a space flight to come by this feeling.' Michael Collins wrote: 'I didn't find God on the Moon, nor has my life changed dramatically in any other basic way.' Astronauts couldn't afford to dream in space, the practical Walter Schirra told Oriana Fallaci. 'I forget my dreams when I'm up there. If I dream I get lost in wonder at the sight of a sunset, at color, I waste the flight and maybe my life.' After discussing the impact of space flight with 'just about everyone who went to the Moon', the shuttle astronaut Don Lind concluded: 'The people who had a profound religious background before they left were impressed in those terms, and those who were too busy to be religious before they left were too busy to be religious before they went back. So I don't think that sort of thing changed anybody.'[37]

Those who were most affected were those who had most time to think: Russell Schweickart, left floating outside his capsule for five minutes thanks to technical problems; Bill Anders, the lunar module pilot with no lunar module; and Gene Cernan, who went to the Moon twice. An interesting explanation has been offered for this variety of experience. Dave Scott called it the 'left seat, right seat hypothesis', after the seating arrangement for the commander and the lunar module pilot. James Irwin reflected:

It is not an accident that the lives of the lunar module pilots have been more changed by the Apollo flights than the lives of the commanders or the command module pilots. The people in my slot were sort of tourists on these flights. They monitored systems that were, for the most part, not associated with control of the vehicle, so they had more time to look out the windows, to register what they saw and felt, and to absorb it.

Edgar Mitchell called it 'the command phenomenon': 'Most of the guys who were vocal about the depth of the experience were lunar module pilots . . . we could take it in and contemplate what we were doing more thoroughly.'[38]

The lives of the lunar module pilots 'tended to follow more unusual, and sometimes difficult, paths than one might expect of pilots and engineers', recalled Dave Scott, citing Aldrin, Bean, Mitchell, Irwin and Duke. The command module pilots who waited in lunar orbit also had time on their hands, among them Michael Collins, Alfred Worden and Stuart Roosa, who told his son how he would look at the Earth gleaming 'like a jewel in the sky', reflect on how it held 'everything I know', and then cover it up with the palm of his hand.[39] It was not just time to think that mattered, but time to think about the Earth.

Returning to Earth was not easy for any of the lunar travellers. 'I had become much more philosophical, at times unable even to focus on minor problems back on Earth,' said Gene Cernan. 'My fellow astronauts who went to the Moon encountered varying degrees of the same disease; we broke the familiar matrix of life and couldn't repair it.' Buzz Aldrin, the second man on the Moon, wrote: 'I travelled to the Moon, but the most significant voyage of my life began when I returned.' The intense competitiveness and workload of astronaut training were succeeded by 'the melancholy of all things done'. In his youth he had had nightmares after reading a science fiction story about a voyage to the Moon in which the travellers returned home insane; now it seemed to be coming true. He went off the rails and underwent psychiatric treatment for depression; his autobiography followed. Edgar Mitchell summed it all up: 'We went to the Moon as technicians, we returned as humanitarians.'[40]

'In their books,' noticed one reviewer of astronaut autobiographies, 'they want, with Aldrin, to stand up and be counted as ordinary people with ordinary lives and ordinary problems.' Russell Schweickart explained about his astronaut colleagues: 'They're not heroes out of books – they're next-door neighbours. Their children and your children play, and they're out there around the Moon reading from the Bible in a way that you know means a great deal to them.' When she was five, Gene Cernan's granddaughter realised that her grandfather was an astronaut: 'I didn't know you went to heaven,' she said.[41]

NASA too eventually adopted a more rounded appreciation of its own astronauts. The turning point was a crew rebellion on Skylab 3 in 1973–4. After the Apollo programme was cut short, a surplus Saturn V booster had been converted into an orbiting laboratory, and three crews went up there in 1973–4, the first two led by the Apollo veterans Pete Conrad and Al Bean. The third mission, however, was crewed entirely by first-time astronauts, and lasted nearly three months. Although the crew had a huge amount of space to float around in, the windows were small and badly positioned, the viewing equipment was in the wrong place, and the binoculars didn't focus properly. They were subject to a relentless regime of technical routines, micro-managed by insensitive flight controllers schooled on the high-intensity Apollo missions. As the *New Yorker*'s science writer Henry Cooper commented, 'It was, somehow, typical of NASA to send men off to a totally new sort of experience, and then overplan it to such an extent that they had no time to think about what they were experiencing – an important reason for sending human beings into space in the first place.'[42]

The crew eventually went on strike and were finally granted more time to themselves, which they spent looking out of the biggest window. They crowded round to watch the sunrises and sunsets – sixteen a day – and they marvelled at the vast uninhabited expanses of Asia and Africa. 'Not much of Earth is hospitable to man,' observed Gerald Carr. 'We don't occupy much of our world. We're crowded into small areas.' 'Every pass was different,' said William Pogue. 'It was never the same from orbit to orbit. The Earth was dynamic; snow would fall, rain would fall – you could never depend on freezing any image in your mind.' 'I gained a whole new feeling for the world,' recalled the science pilot Edward Gibson. 'It's God's creation out before us, and whether you're looking at a bit of it through a microscope, or most of it from space, you still have to see it to appreciate it.' The spacewalks were even better, explained Jack Lousma of Skylab 2. 'When you're inside looking out the window, the Earth's impressive, but it's like being inside a train. . . . But if you stand outdoors, on the workshop, it's like being on the front end of a locomotive as it's going down the track! But there's no noise, no vibration; everything's silent and motionless.' Such experiences, commented Cooper, 'convinced all nine of the Skylab astronauts that the Earth had to be observed directly, as any living object should be, with all the flexibility and intelligence that man could provide, rather than indirectly, by an unmanned satellite, as though it were dead and static.'[43]

NASA learnt its lesson too. At first the flight controllers, conditioned by Apollo, felt that time in space was too valuable to allow the astronauts to have time off, but eventually, one explained, 'I saw they needed time to think about what they were doing and to re-establish themselves. . . . We now feel that an astronaut's time off should be inviolate.' In the 1980s, once it found room to send non-astronauts up in the space shuttle, NASA indicated a preference for passengers who could 'communicate the experience'.[44] For an organisation which depended for its future budget on communicating with American voters, it was a belated realisation.

A number of the astronauts who had been most affected by their experiences in space went on to found the Association of Space Explorers, open to everyone who had been in Earth orbit. Among the founders were Russell Schweickart, Michael Collins, Edgar Mitchell, Oleg Makarov and Alexei Leonov, the first man to walk in space. After two years of careful negotiations, the first working meeting took place near Moscow in 1983, and the Association's first congress was held in France two years later. Leonov wrote the press release. 'Astronauts and cosmonauts are the handful of people who have had the good fortune to see the Earth from afar and to realize how tiny and fragile it is,' he said. 'We hope that all the peoples of Earth can understand this.'[45] The aim of the first congress was 'to protect and conserve the Earth's environment'; its theme was 'The planet, our home'. The conference logo, designed by Alexei Leonov, was a space helmet with the Earth reflected in it; the souvenir poster showed the 'Blue marble' Earth, embellished with all the astronauts' signatures. 'We hope that everyone will come to share our particular cosmic perception of the world and our desire to unite all the peoples of the Earth in the task of safeguarding our common and only, fragile and beautiful home,' wrote Makarov.

The Association gave a special prize to the undersea explorer Jacques Cousteau for his dedication to 'the riches of the natural world'. With support from Edgar Mitchell's Institute of Noetic Sciences, it went on to produce a magnificent large-format book of Earth photography called *Home Planet*, embellished with comments from numerous space explorers. In his introduction, Schweickart wrote: 'It is this shared personal impression of our home planet that has brought many of us together as the Association of Space Explorers. . . . It is the golden thread that connects us all. . . . It is what I ponder now, and what I will marvel over for the rest of my life.'[46]

All of the astronauts and cosmonauts who saw the Earth from orbit knew, more surely than those below, that we all live in one world. But those who saw it from afar understood something more, for they had seen the whole Earth. The ambition to reach out to all humankind cannot have seemed so remote for those who had actually seen the whole Earth at one glance. As they sought to involve the rest of humanity in the experience of space, the first 'envoys of mankind in outer space' became the first citizens of Earth.

CHAPTER 7

From Cold War to open skies

The Apollo photographs of the whole Earth seem to almost everyone who sees them to have some deeper resonance, hard to put into words. They are about more than just space travel and human technological achievement; there is a sense of insight that transcends present-day problems and values. Like all apparently transcendent visions, however, this one has its own historical context.

The period since the end of the Second World War had been filled largely by the Cold War, whose bipolar world and politics of fear had displaced the internationalist ideals which had accompanied the defeat of Nazism and the founding of the United Nations. In Jay Winter's account, 1948 was one of the twentieth century's 'utopian moments', symbolised by the UN Universal Declaration of Human Rights; so too was 1968, with its international wave of liberation protest. Such moments, argues Winter, are the result of a human tendency 'to dream dreams, to erect new edifices, to imagine futures precisely at the moment when those dreams, structures and futures are least likely to be realized'.[1] Below the political permafrost of the Cold War, as during the mayhem of the Second World War, optimistic visions were nurtured and pursued, ready to break out like alpine plants at the first sign of a thaw.

This chapter will look at three areas where new ideas – minor utopias, even – developed in dialogue with the first space age: Earth observation, international law and religion. All three of these fields had their own ideas of the Earth, and at the end of 1968 all three shared in a kind of visual revelation when Apollo 8 offered a sight of an Earth whose wholeness had been all but forgotten amid all of its post-war divisions.

The Earth race

NASA did not notice the Earth, at first. This was odd, since the Earth was what the space programme had originally been about, before all this had been forgotten in the race to the Moon. Satellites, like the atmospheric sounding rockets that were tested alongside the V-2s, were a technology for getting to know the Earth better. The Soviet satellite Sputnik had been launched as part of the International Geophysical Year (IGY) of 1957–8. IGY had begun life as the third 'international polar year' until its brief was expanded. It has been described as 'a grand scheme in a world that was still recovering from a devastating world war' and as 'a defining moment for twentieth-century globalism'.[2]

Space exploration and polar exploration had more in common than is often realised. Men in protective clothing, taking all their supplies with them, explored hostile territory in a remote, alien environment, in the service of human knowledge and under-standing. They also required some of the same technology and the same international cooperation. Both also yielded insights not just into the territory they explored but into the nature of the Earth as a whole.

The Arctic and the Antarctic had two quite different histories in the twentieth century – they were, quite literally, poles apart. The Siberian Arctic was Russia's frontier. Whereas for Americans the open frontier stood for freedom, opportunity and plenty, for the Russians it signified danger, temptation and hardship. In between the wars, Arctic explorers received the kind of press that would later be accorded to cosmonauts as Soviet culture celebrated the victories of human will and technology over 'the cruel Arctic'.[3] On the western side, in the 1940s azimuthal map projections, looking down on the pole, suggested the Arctic as a kind of frozen Mediterranean, surrounded by rival land empires – like the Christian and Islamic worlds, perhaps, or the empires of Rome and Carthage. After 1945 the Arctic became militarised as the nuclear missile and defence systems of the American and Soviet empires confronted each other across the North Pole. It was the coldest theatre of the Cold War. Here, Richard Underwood practised the missile tracking and aerial photography techniques that would eventually produce photographs of the Earth from space. The militarisation of the North Pole dictated that most polar research would be carried out in the south.

While war was planned across the Arctic, peace broke out in the Antarctic. A series of international conferences and panels in the mid-1950s planned Antarctic expeditions and bases, tests of the upper atmosphere and the Earth's magnetic field at the pole, and data-gathering exercises coordinated on the same days across the world. The Antarctic was so inhospitable it was thought to be more like space than Earth; Wernher von Braun visited it. America and Russia discussed satellite launches as part of the IGY programme to investigate the atmosphere, and the Soviets indicated an intention to attempt one, but until it happened no one believed they were serious. The British Antarctic Survey of 1955–7 produced deep core ice samples which were later to provide crucial evidence of past climate change. At the end of the whole process, the Antarctic treaties of 1959–61 declared it to be a global wilderness. Nuclear explosions were prohibited and nations suspended territorial claims in favour of international cooperation and scientific research, prefiguring the Nuclear Test Ban and Outer Space treaties of 1963–8. The Antarctic treaties have been described as 'the first memorable thaw in the cold war'.[4]

Focused from the start on the Moon, NASA had been slow to notice the Earth. For several years the astrofuturist vision of space as a highway to other worlds swept all before it. Funding for many areas of American science suffered as the space programme scooped the pool, although atmospheric research did very well out of it. When in 1965 NASA published a book on *Space Exploration: Why and How* for the general public it presented the value of space photography for Earth entirely in terms of scientific study, giving pride of place to stunning orbital landscapes from the Gemini missions; there was no hint of what it might mean to see the whole Earth, or even that such a thing was in prospect. Unfortunately NASA, effective at organising itself to get to the Moon, turned out not to be so good at working with other agencies for more Earthly goals. For some years NASA still assumed that Earth resources satellites would need to be manned by astronaut observers; only in 1966 did it accept that they could be automated orbital laboratories.[5]

In the mid-1960s NASA began emphasising the scientific benefits of Earth observation as it defended its massive budget against increasing pressure in Congress. Successes with weather satellites, and the selective release of data from classified military satellites, created a demand for information from space. The US Commission on Marine Science chose the welcome ceremonies for the returning Apollo 8 crew to renew the proposal for a 'wet NASA' to invest in the long-term study and development of the oceans. The proposal had considerable support, including that of Senator Joseph Karth, the chair of the Space Science Committee, encouraging NASA to develop its own Natural Resources programme for Earth photography.[6] The eventual result was the Earth Resources Satellite project, later renamed 'Landsat'. Five satellites were sent up between 1972 and 1984, catching the environmental wave that followed the Apollo project. They were used for resource management, hydrology, agriculture, geology and geography, bringing NASA into partnership with large numbers of companies, agencies, universities and governments whose concerns were Earth-centred.

Apollo 8 helped this process along. When the crew of Apollo 8 addressed Congress on their return to Earth, Frank Borman hoped that 'in a few years we will have an international community of exploration and research [on the Moon], much the way we have in Antarctica'. Later that year, on a visit to Russia, he looked forward to a joint Soviet–American space station, run 'much the way we work in the Antarctic'. A useful link here was Simon Bourgin, the Science Policy Officer at the US Information Agency who had helped Borman prepare the Genesis broadcast. His job included both acting as political escort to the returning astronauts and publicising the US role in the Antarctic through the world's media. He went on to cover international environmental conferences, including the 1972 UN Conference on the Human Environment in Stockholm.[7]

Among the unexpected enthusiasts for space was the oceanographer and undersea explorer Jacques Cousteau. Traditionally oceanographers and NASA did not mix, for ocean exploration had lost out to the space programme in the battle for federal funds in the 1960s; calls for a 'wet NASA' usually implied criticism of space as a priority. Cousteau, however, had always thought space travel and deep sea diving had a lot in common. At Christmas 1975, NASA's deputy administrator George Low joined him for a week's diving at an island off New Mexico. The result of the encounter was one of NASA's earliest environmental collaborations, using the new Earth observation satellite Landsat to observe and measure ocean currents and depths on an unprecedented scale.[8]

Ironically, just as the achievements of IGY in mapping the Antarctic were completed, physical exploration was succeeded as a mapping technology by orbital observation. Most of NASA's spacecraft and satellites, however, were of limited use at the poles, for they were in broadly equatorial rather than polar orbits – a side-effect of the objective of sending men to the Moon, and of the demand for weather and communications data from the developed northern hemisphere. As luck would have it, however, Apollo 17 blasted off near the middle of winter, with the South Pole tipped up towards the Sun; as a result, the 'Blue marble' photograph provided the first good view of the Antarctic, almost in its entirety.[9] It seemed appropriate, then, that in the 1980s space technology was instrumental in the discovery of the ozone hole over the Antarctic, sounding an early alarm about global climate change. It also gave NASA, at last, its theme for the 1990s: 'mission to Earth'.

The freedom of space

The reputation of the achievements of the first space age suffered a sharp drop after it was all over. The first distanced, critical histories began appearing in the mid-1980s, and they tended to be sceptical. NASA may have been set up as a civilian agency but, writes Walter McDougall, 'the space programme was a paramilitary operation in the Cold War, no matter who ran it'. The early astronauts were all military test pilots, the early rocket boosters were versions of military missiles with humans in the nose cone and (despite the mythical non-stick pans) the technological spin-offs were mostly military too. In this interpretation, had the Soviet Union not panicked the United States by taking an early lead then nothing so preposterous as the race to the Moon would ever have come about. It was a scientific dead end, a case of pork-barrel politics at work behind a veil of high principle. The Moon itself was 'simply a battlefield in the Cold

War', and once the excitement of the journey was over it turned out to be so boring that no one has been back there since. All this is summed up in a simple phrase: 'the dark side of the Moon'.[10] The Moon, however, has two sides. When its dark side faces the Earth, the Earth itself is fully lit – exactly how the 'Blue marble' photograph was taken. On the dark side of the Moon lay the Earth.

The space programme was motivated by some of the darkest fears of the twentieth century but it also carried some of its most extravagant dreams. The promise of space was constantly counterposed to the threat of nuclear war, almost as if the two were mutually exclusive – which in a way they were, since nuclear war would have closed off the high-tech future in which the space programme was expected to operate. The race to the Moon acted as a displacement of the nuclear arms race – perhaps a decisive one, although such things can never be known. As Jay Winter has written, the emergence of total war in the twentieth century was a stimulus to utopian ideas; there was a 'complex and subtle dialogue between minor utopian visions and massive collective violence'.

Astronauts were well placed to appreciate this paradox. As they trained for space, their fighter-pilot former colleagues were flying tours of duty in Vietnam; 'they fought my war,' felt Gene Cernan. Frank Borman had served in Berlin during the early Cold War. In the 1950s Buzz Aldrin flew with the US Air Force in Germany. He was reassured to be told that 'our targets would be military and not civilian', but 'because many of our actual targets would be behind the iron curtain, we would take off and plan our mission in such a way that if it were the real thing we'd drop our bombs and head for a neutral country. The presumption was that Germany would be either occupied or destroyed by the time we finished our missions.' As a test pilot in the same period, Edgar Mitchell became involved in developing delivery systems for nuclear bombs; disturbed at what he was doing, he set about acquiring the scientific training to become an astronaut, and he eventually succeeded.[11] The astronauts who beheld the whole Earth had more reason than most to understand the fragility of what they were seeing; the transformation of some into apostles of the unity of humankind was entirely understandable.

The particular paradox of the age was that military competition helped to establish the freedom of space. The combination of arms race and space race forced the development of an international law governing space, which in turn meant that Apollo 8's photograph of the Earth appeared in a very different context than did the first manned Mercury flight at the height of the Cold War. The 1950s saw horrifyingly rapid developments in both the destructiveness of nuclear weapons and the willingness to use them, a willingness born of fear of what the enemy might do first. Eisenhower, the warrior president (1953–61) who gained first power and then wisdom, was haunted by the memory of Pearl Harbor and the fear of a surprise attack. But he also saw that intelligence about what the Soviet Union was up to was the key to a sane (and affordable) strategy. The U-2 spy planes which overflew the Soviet Union from 1956 were one solution, but they were both illegal and highly risky; one was shot down in 1960, rupturing relations between the superpowers. The future, Eisenhower realised, lay with satellite reconnaissance.

While long-distance aeroplanes and missiles had developed as weapons of war, satellites mostly carried an aura of peace. When an American satellite launch was proposed as part of the International Geophysical Year of 1957–8, Eisenhower had the National Security Council secretly look into the plan. It reported that 'a small scientific satellite will provide a test of the principle of "freedom of space"', offering 'an early opportunity to establish a precedent for distinguishing between "national air" and "international space", a distinction which could be to our advantage at some future date when we might employ larger satellites for intelligence purposes'. The CIA warned that whoever launched the first satellite would gain 'incalculable prestige and recognition throughout the world'. If it was the USSR, it would appear to provide 'sensational and convincing evidence of Soviet superiority'. A US launch would have to be done 'with unquestionable peaceful intent . . . . To minimize the effectiveness of Soviet accusations, the satellite should be launched in an atmosphere of international goodwill and common scientific interest.'[12]

Eisenhower threw his weight behind the satellite programme and over the next few years repeatedly challenged the Soviet Union to agree to an 'open skies' policy, using mutual aerial surveillance as a guarantee of peace. The Soviets resisted, for their military credibility depended upon bluff; they were still far from being able to bomb the

US. But they quietly prepared to launch the first satellite themselves, under the banner of the International Geophysical Year. When Sputnik 1 went up in October 1957, the USSR took care to remind everybody of the IGY, while the US forgot about it completely. There was panic: if the Soviets could put a satellite in orbit, could they do the same with a nuclear bomb? One newspaper headline summed up the fears: 'Pearl Harbor in space'.[13]

Eisenhower resisted being bounced into an expensive arms race in space. He understood, as his defense secretary put it, that 'the Russians have . . . done us a good turn, unintentionally, in establishing the concept of freedom of international space'. Instead of protesting about Sputnik's violation of American airspace, he hastened to send American satellites over Soviet airspace. In 1958 NASA was set up as a civilian agency, closely involved with military projects and contractors but formally distinct from them in its mission. The 1958 National Aeronautics and Space Act stated: 'It is the policy of the United States that activities in space should be devoted to peaceful purposes for the benefit of all mankind.' At the same time the United Nations (including Russia) set up a Committee on the Peaceful Uses of Outer Space. It declared in favour of the freedom of space and resolved that 'outer space should be used for peaceful purposes only'.[14]

John F. Kennedy won the presidential election of 1960 for the Democrats by capitalising upon fears that the Soviets had built up a dangerous lead in the nuclear arms race – the notorious 'missile gap'. 'If the Soviets control space they can control Earth', he warned, following Wernher von Braun. 'To ensure peace and freedom, we must be first.' In fact the US was way ahead in missiles and the American public had (indirectly) fallen victim to Soviet propaganda. Eisenhower's administration realised that 'the most significant factor of Soviet space accomplishments is that they have produced new credibility for Soviet statements and claims',[15] but Eisenhower could not reveal what he knew about true Soviet capabilities without revealing the secret programme of U-2 spy flights.

As President, Kennedy had a similar problem. He could not reveal that the United States was winning the missile race without revealing that the 'missile gap' had been a hoax all along. Nor did he want to draw attention to the fact that the latest concepts of nuclear superiority were based on a terrifying strategic theology of nuclear apocalypse. In any case, the myth of Soviet nuclear superiority was to remain useful for as long as the Cold War lasted.[16] The US could not admit that it was winning the arms race, but winning the space race was another matter.

Kennedy's carefully chosen finishing line for the space race was announced in May 1961: man on the Moon by the end of the decade. It seemed to lift the whole enterprise onto a new level. In February 1962 America symbolically drew level in the space race by sending John Glenn into orbit, an event that was greeted with national rejoicing on a scale not seen since 1945. Khrushchev immediately offered cooperation in space exploration, and Kennedy publicly responded the next day by commenting: 'We believe that when men reach beyond this planet they should leave their national differences behind them.' If such cooperation could be achieved, he suggested to an audience at the University of California, 'the stale and sterile dogmas of the Cold War could be literally left a quarter of a million miles behind'. A few weeks before his assassination, Kennedy proposed to the United Nations a joint US–Soviet mission to the Moon in which the astronauts would be 'not the representatives of a single nation, but the representatives of all countries'. In the end they were both, leaving behind both the stars and stripes and a plaque proclaiming 'we came in peace for all mankind'.[17]

Kennedy's goal was not simply to beat the Russians, or even to appear to be racing them, but to do something spectacular to rise above it all. The space programme served the purpose precisely because it was identified with the universal values of human progress, not the narrow values of national competition. It was both, of course, but it could only serve the narrow purpose because it appeared to serve the wider one. While in the short term the Sputnik panic intensified both the arms race and the space race, in the medium term there followed important international treaties about space, and then the Apollo programme. In the depths of the Cold War, peace broke out in space.

The pivotal event in creating the freedom of space was the Cuban Missile Crisis of autumn 1962. As the United States resisted an attempt by the Soviet Union to place nuclear missiles within range of its territory, the world came to the brink of nuclear war.

Like nothing before or since, this crisis focused minds on the future of humanity. The concept of the 'nuclear dilemma' became established in the public consciousness. The dilemma, in the original sense of a fork-in-the-road decision between two alternatives, lay between indefinite progress and nuclear extinction. Kennedy told the students of Rice University, Texas, that the quality of American leadership in space would determine 'whether this new ocean will be a sea of peace or a terrifying theater of war'. As an early treatise on space law put it, 'the statesmen of the contemporary Earth–space community' had the power either 'to smooth the way to an optimum order of undreamed-of abundance and benevolence' or 'by default . . . by their timidities and mistakes they can end history – as man records it'. Here was the optimism of the postwar years projected into the future.[18] Thanks to the efforts of the space publicists of the 1950s, the peace option was inextricably associated with space travel and the astrofuturist ideal. In seeking to avoid nuclear extinction, statesmen turned instinctively to space.

The first formal treaty governing space was a direct result of the Cuba crisis. The Nuclear Test Ban Treaty of 1963 (also known as the Partial Test Ban Treaty) banned nuclear explosions of any kind 'in the atmosphere; beyond its limits, including outer space; or under water'. The Orion programme, America's project to propel a rocket into space by nuclear explosions, was quietly wound up. The same year, the United Nations General Assembly resolved to ban weapons of mass destruction from 'outer space and celestial bodies', including the Moon.[19] Nuclear war was to be kept out of space and thus symbolically out of the future. There were (at least until the 'Star Wars' programme of the mid-1980s) to be no more nightmares of nuclear war being waged from beyond the Earth.

In the mid-1960s space calmed down in another way. The first 'man in a can' phase of manned spaceflight, all urgency and nose cones and rocket exhausts, was over. The Gemini programme of 1965–6 by contrast seemed graceful and unhurried. These years saw a steady succession of two- and three-manned missions, performing stately orbital manoeuvres and spacewalks and sending back stunning pictures of the Earth below. The Gemini pictures seemed like evidence that humanity had at last risen above its Earthly fate to see its own serene future.

With space travel apparently almost routine, in January 1967 the United Nations issued the Outer Space Treaty, the product of several years of negotiation. Its first article read:

> The exploration and use of outer space, including the Moon and other celestial bodies, shall be carried out for the benefit and in the interests of all countries, irrespective of their degree of economic or scientific development, and shall be the province of all mankind.

There was to be no appropriation, no claim of national sovereignty and no stationing of weapons of mass destruction in outer space; the Moon and other celestial bodies were to be used 'exclusively for peaceful purposes'. Outer space was to be a limitless international zone of freedom, equality and scientific cooperation. Astronauts were recognised as the 'envoys of mankind in outer space'. This echoed the words of the 1961 Antarctic Treaty which neutralised Antarctica for scientific exploration 'in the interest of all mankind'. It also drew on the principles of the 1962 Law of the Sea Convention, which prevented nations from appropriating the high seas. In the end, the Antarctic and Outer Space treaties were both the result of the International Geophysical Year, which set in train both polar expeditions and the first satellite launches.[20] After only a decade of exploration, and without a single permanent resident, space had a constitution.

The 1967 Outer Space Treaty was not without its loopholes and limitations. While nuclear weapons could not be 'stationed' in space, they might travel through it – so intercontinental ballistic missiles were OK provided they didn't actually orbit. 'Peaceful purposes' were generally agreed to include surveillance, defence of one's own satellites and perhaps even 'non-aggressive' military preparations. But while space was not totally demilitarised, 'celestial bodies', including the Moon, were. 'Outer space' was not defined, although by general consent it began in orbit, with everything below counting as 'airspace' under national sovereignty.[21] None of these problems proved serious in practice. The 1967 treaty was followed by the 1968 Space Rescue Treaty, which obliged all nations to help out any astronauts in difficulties, in the same way as if they were sailors on the

Colour copies of some of these photographs, and of others which cannot be reproduced here, can be seen at the Earthrise website, www.earthrise.org.uk

1 Earthrise, seen for the first time by human eyes, 24 December 1968. This was how the crew of Apollo 8 saw the Earth appear from the side of the Moon, with the north pole at the top.

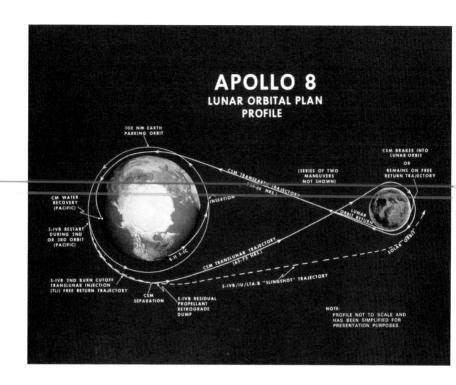

APOLLO 8
LUNAR ORBITAL PLAN
PROFILE

100 NM EARTH
PARKING ORBIT

CM WATER
RECOVERY
(PACIFIC)

S-IVB RESTART
DURING 2ND
OR 3RD ORBIT
(PACIFIC)

S-IVB 2ND BURN CUTOFF
TRANSLUNAR INJECTION
(TLI) FREE RETURN TRAJECTORY

CSM
SEPARATION

S-IVB RESIDUAL
PROPELLANT
RETROGRADE
DUMP

S-II S-IC

INSERTION

CSM TRANSEART': TRAJECTORY
22-60 HRS.

(SERIES OF TWO
MANEUVERS
NOT SHOWN)

CSM BRAKES INTO
LUNAR ORBIT
OR
REMAINS ON FREE
RETURN TRAJECTORY

LUNAR
ORBIT RETURN

CSM TRANSLUNAR TRAJECTORY
(65-75 HRS.)

S-IVB/IU/LTA-B "SLINGSHOT" TRAJECTORY

SOLAR ORBIT

NOTE:
PROFILE NOT TO SCALE AND
HAS BEEN SIMPLIFIED FOR
PRESENTATION PURPOSES.

2 The plan for the first voyage beyond the Earth. The Earth and the Moon are imagined as quite similar – very different from what Apollo 8 was to reveal.

3 'There she is, floating!' Apollo 8 transmitted the first live pictures of the whole Earth at Christmas 1968. This was the view as the crew returned home.

4 Frank Borman's Earthrise. Taken just before the more famous photo, it was overlooked for nearly thirty years because it was in black and white.

5 Presenting Earthrise. President Kennedy did not live to see the Earth, but Texas Governor John Connally, injured in the same assassination, did. Here he receives a framed Earthrise from the crew of Apollo 8.

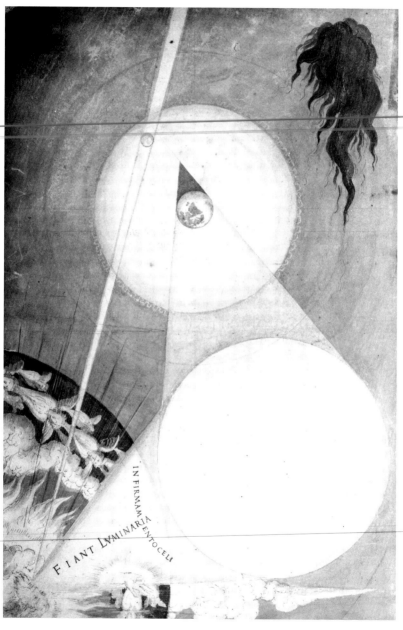

Text within the image: FIANT LVMINARIA IN FIRMAMENTO CELI

6 The creation of the Sun and Moon by Francisco d'Olanda, 1547. An uncanny precursor of the sight of the Earth from space; in the original, the Earth is blue and white.

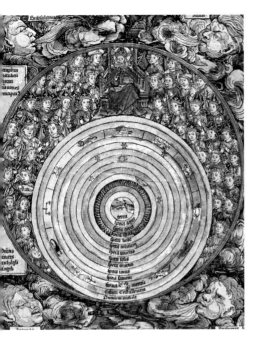

7 The late medieval cosmos. A lowly Earth, naturally at rest in the centre of the cosmos, circled in ascending order by the Sun, Moon, planets and stars, the whole closely supervised by God and his angels just beyond. Earth is unique and the universe is very limited in both time and space. From Hartman Schedel, *Liber Chronicarum*, 1493.

The first distant image of the Earth, transmitted by the Explorer VI satellite in August 1959. 'Scientists can discern cloud banks in the large white areas,' announced NASA.

9 A V-2 rocket launch at White Sands, New Mexico.

10 The curving horizon, photographed from about 45 miles up by a V-2 on 26 July 1948.

V-2 ROCKET-EYE VIEW FROM 60 MILES UP

1- MEXICO	3- LORDSBURG, NEW MEXICO	6- SAN CARLOS RESERVOIR	9- SAN MATEO MTS.	12- ALBUQUERQUE · NEW MEXIC
2- GULF OF CALIFORNIA	4- PELONCILLO MTS.	7- MOGOLLON MTS.	10- MAGADALENA MTS.	13- SANDIA MTS.
	5- GILA RIVER	8- BLACK RANGE	11- MT. TAYLOR	14- VALLE GRANDE MTS.
				15- RIO GRANDE
				16- SANGRE DE CRISTO RANGE

ROCKET FIRED AT WHITE SANDS PROVING GROUND, JULY 26,1948   DISTANCE FROM CAMERA TO HORIZON-700 MILES

AREA SHOWN APPROXIMATELY 800,000 SQ.MILES   DISTANCE ALONG HORIZON-2700 MILES   INSTRUMENTATION AND PHOTOGRAPHY BY APPLIED PHYSICS LAB THE JOHNS HOPKINS UNIVERSITY FOR THE BUREAU OF ORDNANCE

11 'Columbus was right!' A mosaic panorama stretching from the Pacific to the Rio Grande assembled from photographs taken 60 miles up on the V-2 flight of 26 July 1948.

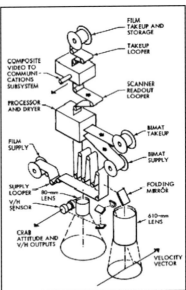

12 The Soviet Earthrise. This little-known photograph, in black and white, was brought back to Earth by the unmanned Zond 6 mission in November 1968, beating Apollo 8.

13 The ingenious orbiting photo laboratory used by Lunar Orbiter to take, process, scan and transmit the first photo of the Earth from the Moon in August 1966.

14 Earthrise as seen by the Lunar Orbiter probe, August 1966. No such shot was on the mission schedule: was it a response to Stewart Brand's campaign for a picture of the whole Earth?

15 'The whole world in his hands'. An engineer from the Hughes Aircraft Corporation holds aloft the ATS-III whole Earth photograph.

16 A day on Earth. The Earth waxes and wanes like the Moon, 18 November 1967, as seen from the ATS-III satellite.

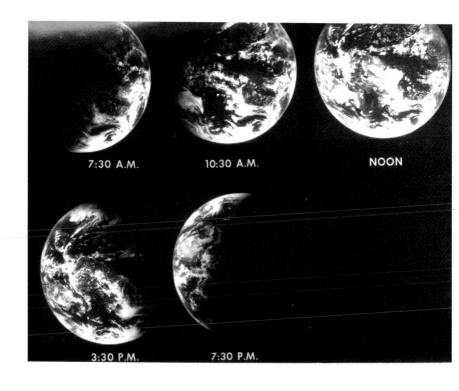

7:30 A.M.　　10:30 A.M.　　NOON

3:30 P.M.　　7:30 P.M.

17 The whole Earth at last. The first good-quality picture of the whole Earth, taken in colour from 22,300 miles up by the ATS-III satellite, November 1967. Parts of the Americas, Africa and Europe are all visible.

18 The 'Blue marble', taken by Harrison Schmitt on the way to the Moon aboard Apollo 17 in December 1972. A photographic manifesto for global justice, and the single most reproduced image in human history.

19 For some of those who walked on the Moon, the height of the experience was to view the Earth. 'One Small Step for a Man' by NASA artist Alan B. Chinchar.

20 The Earth was not easy to see from inside a bulky spacesuit on the Moon, as shown in this photograph of Harrison Schmitt by Gene Cernan.

21 'The saddest day of my life'. Ed White could hardly bear to return to the capsule after his spacewalk on Gemini 4, June 1965. His friend Russell Schweickart later wrote a meditation after his own spacewalk on Apollo 9, entitled 'No Frames, No Boundaries'.

22 The Earth, signed by the photographers: the poster for the first conference of the Association of Space Explorers, 1985.

23 'In the beginning, God…'. The US Postal Service commemorated Apollo 8 with religious wording that was unusual for the period.

24 The Genesis petition. Loretta Lee Frye of the Apollo Prayer League presents NASA chief Thomas Paine with half a million signatures in support of the Apollo 8 astronauts' reading of the Bible from space.

Why haven't we seen a photograph of the whole Earth yet ?

25 Stewart Brand's campaign badge, February 1966.

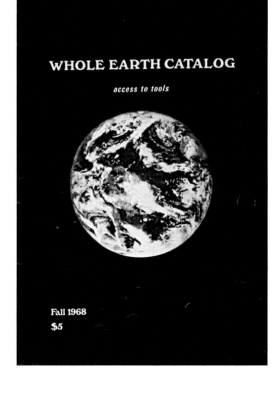

26 The first *Whole Earth Catalog*, autumn 1968, featuring the ATS-III photograph of the Earth.

27 John McConnell and Rena Hanson display the Earth flag for the first time during the Moonwatch at New York's Central Park in July 1969.

28 A button badge, from the first Earth Day in 1970.

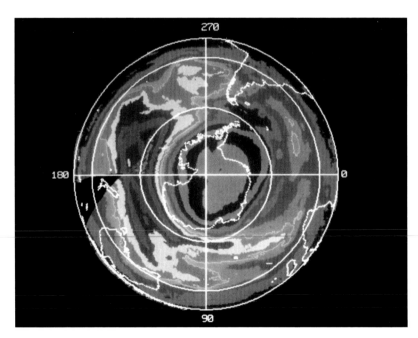

29 The Earth hole: the Antarctic ozone hole, seen from space in October 1986, was one of the first signs of human impact on a planetary scale.

30 Deep perspective: the Earth and Moon, taken by the Galileo spacecraft from four million miles out, December 1992. A similar photograph had been taken by the departing Voyager probe in 1977.

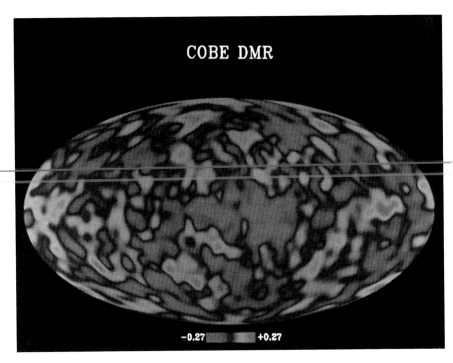

31 The whole universe. The cosmic microwave background radiation (the 'echo of the big bang'), mapped in 1992 by NASA's COBE telescope. Earth is at the centre.

32 Japan's Earthrise. In April 2008 the Japanese lunar probe Kaguya (Selene) took a high-definition digital film of the full Earth rising over the Moon. The probe was coming round from below the Moon at the time, and Earth's south pole is at the top.

high seas. As it came into force on 1 December 1968, the Apollo 8 astronauts were the first potential beneficiaries.

Behind some of the commentaries on the meaning of Apollo 8 lay the shadow of nuclear war. Shortly before it set off, the *New York Times* warned that there would be no room in space for 'wasteful rivalry deriving from Earthbound nationalistic and political ambitions . . . the only rational response is cooperation to make space an arena of unity and international brotherhood'. After these 'Columbuses of space', thought the paper, should come not conquistadors but 'ships bearing the United Nations flag, each carrying men of different citizenship, language, political and religious convictions and color'. The *Boston Globe* warned that 'the light shed on Earth by Apollo 8 was a light that can be dimmed quickly if we do not also see, in its disclosure of the smallness of our planet, how infinitely tawdry are the differences that separate its races, its nations, its men from one another'. Thinking of the Earthrise photo, Bill Anders said: 'it showed the political leaders that there really was only about that much difference between Washington and Moscow . . . all of the views of the Earth from the Moon have let the human race, and its political leaders, and environmental leaders, and its citizenry realize that we're all jammed together on one really kind of dinky little planet, and we better treat it and ourselves better, or we're not going to be here very long.'[22]

Internationalism received a boost from Apollo 8. The *Christian Science Monitor* proposed making the Moon a zone of international science, like Antarctica, 'lifting mankind's sights beyond the ancient rivalries of Earth'. The *Chicago Sun-Times* found the sight of Earth 'awesome', but pondered on the hatred, misunderstanding and grief hidden below the clouds. 'If a single nation can prepare with confidence to place a man on the Moon, what could a world of nations united in the cause of common good accomplish?'[23]

Strangely enough, the Outer Space Treaty was barely noticed in this flood of internationalist brotherhood. One far-sighted stamp dealer sought the signature of NASA's chief, Thomas Paine, on first day covers of the Apollo 8 stamp, printed with the text of the Outer Space Treaty and the Space Rescue Treaty, but that was about all.[24] The ceremonial signing of the Outer Space Treaty nearly two years earlier had been interrupted by the news that three Apollo astronauts had been killed in training in a launch pad fire, which drowned out press coverage of the event. The creation of the freedom of space, however, grew in significance as the years passed. It came into force on the eve of the Apollo programme, making its astronauts the first official 'envoys of mankind'. As the crew of Apollo 8 addressed the Earth from the heavens, their choice of words, speaking to mankind about itself, was politically spot-on. Soon afterwards, their photographs showed mankind its home. The barren lunar landscape struck some as like a battlefield; the rising Earth seemed like peace itself.

Did you see God?

Apollo 8 marked the apex of human technological achievement to date, yet the pictures of Earth which it sent back seemed to speak of the primeval, of origins, of faith. This almost religious impact was in part down to its association with the crew's Christmas Eve reading of the Creation story from the Bible. As they did so the first recognisable TV pictures of Earth, transmitted on 23 December, were on the front pages of the daily newspapers. While the TV screens showed the barren Moon unrolling below, what stuck in minds were Lovell's words from early in the broadcast contrasting the 'awe-inspiring loneliness' of the Moon with the Earth as 'a grand oasis in the vastness of space'. The closing words of the broadcast punched the connection with Earth right home: 'And God saw that it was good . . . God bless all of you on the good Earth.' When the Earthrise photograph appeared soon after splashdown, showing the Earth apparently fresh from creation, the same phrases were quoted in the press all over again, cementing the connection. Frank Borman was not the only one who thought, 'this must be what God sees.'

Ever since Sputnik a question had been floating uneasily around, posed most clearly by the writer C. S. Lewis: 'Will we lose God in Outer Space?' The first (and probably mythical) rocket traveller, Larari Hasan Celebi, had announced that he was 'going to have a talk with the prophet Jesus'.[25] James Lovelock, living in rural Ireland at the time of the first Moon landing, found that 'to many of those living on the remote Beara Peninsula, heaven was still simply up there in the sky and hell beneath their feet'. When they heard the news they came to him to discuss it. 'Their faith was not perturbed by the news of men walking on the Moon, but their religious belief seemed to be undergoing an internal reorganization.'[26] When human travellers entered what had previously been 'the heavens' and sent back photographs of views previously available only to God, they assisted the spread of a more resilient and universalist strain of Christianity. But as the first astronauts prepared to leave Earth for the heavens, at the back of many minds lay a nagging worry: what would they find there?

In August 1962 two cosmonauts, Pavel Popovich and Yuri Gagarin, appeared on Soviet television. The compère put to them a question asked by a seventy-year-old woman in the audience: did you see God when you were in orbit? Both men laughed. 'We didn't see anything or anybody,' said Popovich.[27] The old woman's question may seem naive, but it also showed awareness of recent reports in the Soviet media – reports that were taken no less seriously in the United States.

A few months earlier, another cosmonaut, Titov, had said: 'Sometimes the people say that God is living there [in space], but I haven't found anyone.' He added: 'I don't believe in God. I believe in man – his strength, his possibilities and his reason.' A year before that, when Yuri Gagarin became the first man in space, the state news services triumphantly reported his comment: 'I do not believe in God, I do not believe in talismans, superstitions, and such things.' The first woman in space, Valentina Tereshkova, professed herself unable to understand how the American Jerrie Cobb found prayers relevant to the piloting of supersonic planes. President Khrushchev himself told an American reporter:

> As to paradise we have heard a lot about it from the priests. So we decided to find out for ourselves. First we sent up our explorer, Yuri Gargarin. He circled the globe and found nothing in outer space. It's pitch dark there, he said: no Garden of Eden, nothing like heaven. So we decided to send another. We sent Gherman Titov and told him to fly for a whole day . . . . Well, he took off, came back, and confirmed Gargarin's conclusion. He reported there was nothing there.[28]

There was a whole vein of Russian jokes about astronauts who met God, tending to subvert the official line that heaven was empty.[29]

Soviet statements were designed to sharpen the ideological conflict over space, and they succeeded. Vatican Radio denounced Titov: the cosmonauts may not have seen God but God certainly saw them. Moscow Radio scoffed that all this was an attempt 'to whip up the enthusiasm of believers, which the achievements of science have caused to waver considerably'. The Vatican should launch its own rocket to find God before it criticised the findings of others. Americans joined in. The Sunday after John Glenn's first orbital flight there was thanksgiving in churches across America. The minister of Zion Lutheran Church, Danvers, Illinois preached a sermon on 'Our great God of space': 'even from the Moon, or any other heavenly body,' he proclaimed, 'the wonderful greatness of God could be seen in the beauties of His creation, no matter from what viewpoint.' Space was big, but God was bigger. An outraged senator from South Dakota read a newspaper editorial denouncing Gagarin's statement into the Congressional record, and predicted that the faith of American astronauts would prove more durable than Soviet atheism.[30]

The astronauts, imbued with American Bible culture, were happy to play their part in all this.[31] 'I got on this project because it would probably be the nearest to heaven I will ever get,' said John Glenn. Invited by a Senate committee to respond to Gagarin, Glenn responded with due humility: 'God . . . will be wherever we go.' A few months later X-15 pilot Robert White, who had just become America's first official 'winged astronaut' after flying to a height of nearly sixty miles, said: 'This world of scientific discovery, however vast it is, is small in comparison with God's knowledge. This is what impresses me up there.' Gordon Cooper offered a well-publicised prayer of thanks for the beauties of creation from his orbiting Mercury capsule.[32] Gemini astronaut James McDivitt told a press conference in Rome: 'I did not see God looking into my space cabin window as I did not see God looking into my car's windshield on Earth. But I could recognize his work in the stars as well as when walking among the flowers in a garden. If you can be with God on Earth you can be with God in space as well.'[33]

The outrage of American Christians at the arrogance of Soviet atheism fed on a wider unease (and perhaps guilt) about the ascendancy of technology and materialism in modern life. Wernher von

Braun had offended religious sensibilities in 1957–8 with his assertion that 'space flight will free Man from his remaining chains, the chains of gravity which still tie him to this planet. It will open to him the gates of Heaven.' Space travel, he told Oriana Fallaci, was 'the will of God. . . . If God didn't want it, he would stop us. Yes of course I'm religious.'[34] To some, this kind of talk resembled the folly of the people of Babel, who had been punished by God for attempting to build a tower to heaven.

In 1958, as NASA was founded, the evangelical *National Council Bulletin* published a roundtable discussion on God and the space race, debating issues such as whether space travel would weaken faith, and why the godless Russians were in the lead. The Lutheran theologian Martin J. Heinecken sent a questionnaire to the heads of the major American Protestant Churches, asking: 'Will space explorers discover God and heaven?' All answered 'No.' 'Place concepts are hardly proper when we are thinking in terms of the spirit,' said a Methodist leader. 'God is known by faith, not by exploration,' replied a professor of Christianity. There was general agreement that the 'three-storey universe' of the Bible had gone, to be replaced with a sense of the divine as pervading the cosmos rather than standing outside it.[35] This was exactly the feeling conveyed by Earthrise, a photo of creation taken from within creation, accompanied by a reading of the Creation story from lunar orbit.

Of the three Apollo 8 astronauts, Anders (who took the famous photo) was a Roman Catholic, and Borman and Lovell were Episcopalians – that is, American Anglicans. It was the more fundamentalist Protestant Churches that were most defensive over the threat to religion of space travel, while the older Churches were more relaxed and adaptable. The Roman Catholic Church, used to thinking on the cosmic scale, took the lead. Pope Pius XII had already welcomed the big bang hypothesis: 'It seems that science of today, by going back in one leap millions of centuries, has succeeded in being witness to that primordial Fiat Lux' – the Creation. Earthrise later aroused similar feelings. In 1956 he received the 400 delegates of the International Astronautical Congress at Castel Gandolfo near Rome, assuring them that their efforts were 'legitimate before God'. 'The more we explore into outer space, the nearer we come to the great idea of one family under the mother-father God. God has no intention of setting a limit to the efforts of man to conquer space. . . . Man has to make the effort to put himself in new orientation with God and his universe.' This reorientation was far-reaching. Pius XII's 1963 encyclical *Pacem in Terris* (Peace on Earth) encompassed scientific discoveries and human rights alike in the theme of 'order in the universe'. In 1962 a conclave was held in Rome to work out how to apply theology to extraterrestrial life: 'the church immutably moves towards space theology,' wrote one theologian.[36]

The next Pope, Paul VI, following a long Vatican tradition, was an enthusiast for astronomy. NASA's James Webb discreetly culti-vated good relations, and the two men exchanged carefully crafted messages about the potential of space travel to transcend human conflict and enhance understanding of the universe. In 1965 the Pope received a personal delivery of a film of the Ranger probe's pictures of the Moon from NASA's general manager, praised the achievements of Italian space engineers, blessed the Gemini 6 astro-nauts ('those who are exploring astral paths'), deplored the nuclear arms race, expressed hope that 'those who on Earth appeared divided by unbridgeable distances may . . . meet and collaborate in space', and signed a Mariner photo of Mars, exclaiming 'vidimus et admirati sumus' (we saw it and were amazed). His consistent theme was that space travel was another way of exploring the divine Creation, and that all human achievements were ultimately the work of God.[37]

The papal delegate in Washington DC was invited to the official party at the Apollo 8 launch, and the Pope himself watched the Apollo 8 splashdown on live TV, afterwards addressing the crowds in St Peter's Square; 'the stature of man in prodigious confrontation with the cosmos emerges immensely small and immensely large,' he mused. On receiving Frank Borman at the Vatican during his European tour, the Pope commented that Apollo 8 had 'added to man's knowledge of God's work, thereby including his appreciation of the glory of God, which is manifested in creation'. After the Apollo 11 Moon landing, James Lovell came to present the Pope with one of the first globes of the Moon. Meanwhile, Catholic litur-gists drew up plans for a chapel on the Moon, as part of the lunar base projected by NASA for the end of the century.[38]

Aware that the First Amendment to the US Constitution forbade the promotion of religion by the state, NASA had always been careful to avoid any official endorsement of religion.[39] It was, however, happy for astronauts to make religious comments off their own bat, and for its own speechwriter to assist them in doing so. 'Any mention of travelling in the heavens sent her rushing to her Bible, which had long since been worn thin,' recalled Buzz Aldrin. NASA contrasted its own astronauts' freedom of speech with the assumed requirement of Soviet cosmonauts to follow the Communist Party line. This laid-back approach was carried to extraordinary lengths on Apollo 8. In the run-up to the mission, Commander Borman had a disconcerting interview:

> About six weeks before launch I got a call from Julian Scheer, NASA's deputy administrator for public affairs. 'Look, Frank,' he said, 'we've determined that you'll be circling the Moon on Christmas Eve and we've scheduled one of the television broadcasts from Apollo 8 around that time. We figure more people will be listening to your voice than that of any man in history. So we want you to say something appropriate.'

That was all the advice the press officer gave: 'say something appropriate'. The words of the first astronauts to visit the Moon would seem to have been unscripted. It was, said Borman in 1999, 'another example of the wonderful country we live in'.[40]

The idea of reading from the Book of Genesis took some time to emerge. Questions at the pre-launch press conference on 7 December were dominated by the Christmas timing of the launch, which some suspected was a NASA publicity stunt (in fact, it was all down to the limited monthly launch window), and others thought was irreverent.[41] Lovell cut off a cynical reporter in mid-sentence: 'I can't think of a better religious aspect to the flight than to further explore the heavens.' Anders joked that as a Catholic he had a dispensation, and talked about generating a Christmas feeling 'among all the peoples of the world'. Borman then shifted the ground neatly with a prescient comment:

> I think we are all leaving here with the feeling that was very well expressed at the last Press Conference by a German correspondent that when you're finally up at the Moon looking back at the

Earth, all those differences and nationalistic traits are pretty well going to blend and you're going to get a concept that this is really one world and why the hell can't we learn to live together like decent people.

These first thoughts were discarded before the mission, as Borman explained after the capsule was lifted out of the sea: 'We first thought perhaps . . . it would be more of a one-world theme where we would tell everyone on Earth, that gee whiz, we are all living on one Earth.' 'I personally didn't do it so much as a religious message but as a message to help underscore the significance to mankind of this first away from our home planet flight,' Anders later recalled.[42]

NASA too seems to have expected something along 'one world' lines. Its chief, Thomas Paine, who a few months before had told a magazine reporter how pictures from space 'emphasize the unity of the Earth and the artificialities of political boundaries', predicted on TV the night before the launch that the sight of the Earth from the Moon would make people realise that the world was one.[43] Borman, however, was too busy to think about it much and asked a publicist friend to suggest something: Simon Bourgin, the experienced Science Policy Officer at the US Information Agency. Bourgin in turn asked a journalist, Joe Laitin. Laitin, a Protestant, naturally searched the New Testament, but could find nothing with sufficiently universal appeal. Late at night, his ideas exhausted, he mentioned the problem to his wife, a Roman Catholic, who started looking in the Old Testament and quickly came up with the opening chapter of Genesis: 'Why don't you begin at the beginning?' Laitin recognised the primeval power of this account of the Creation, with its evocative description of the Earth. He added a closing line and advised that the broadcast be followed by a period of silence at Mission Control.[44]

Bourgin included all this in notes which he sent to Borman on 13 December. Stressing first that 'what you have to say has to be *all* Frank Borman', he went on to offer detailed suggestions for the second Christmas Eve broadcast, which, he felt, should be different from the others in tone, and more Earth-centred. The first suggestion was to describe 'how Earth looks from the Moon'. Borman had told a Far Eastern audience that if Mars was the red planet, Earth's waters could make it 'the blue planet' – is it? Bourgin recommended describing the Moon 'with the detachment of a scientist', but his memo was dominated by attempts to anticipate the astronauts' thoughts about the Earth.

> Does the fact that this faraway planet holds all the things that are dear to you have any special impact at the moment? As you gaze at the distant planet Earth, you are aware that at this very moment each of its three-and-one-half billion inhabitants who has any knowledge of your mission – regardless of wealth, race, tongue, culture, national loyalty, politics, or religious affiliations – is thinking of you and your two companions.

He then typed out all the first ten verses from the Book of Genesis, together with the words of the final greeting as they were later used.

After a telephone discussion with Borman on 15 December, Bourgin thought better of Borman's making 'preachy' comments to the whole Earth – 'they could sound forced and artificial.' He should restrict any mention of 'peace' and 'one world' to once each, and above all not try to fit anything in after the Genesis reading – 'you can't top the Bible.' So the final broadcast seems to have been the result of a genuine dialogue between Borman and his adviser, with Borman considering and rejecting 'one world' sentiments but adopting the Genesis idea to the letter. The crew thereafter fended off questions about their historic broadcast with serene confidence; only a few senior NASA officials knew what they were planning to say.

NASA's trust in its astronauts was superbly vindicated. As one NASA administrator commented, 'the choice of that passage . . . could not have been more perfect . . . America has this inventory of people and here come the people when you need them; out they come . . .' For James Lovell, as for Anders, the point of reading from Genesis was not narrowly Christian; rather, the Creation story was 'the basis of many of the world's religions'. Even the secularly inclined BBC space correspondent Reginald Turnill felt that the reading 'struck one instantly as a stroke of genius'.[45] For most hearers, awed by the display of technological power and struggling to decode space jargon and strange TV pictures of the lunar landscape, it was profoundly comforting that an out-of-this-world experience could be expressed in familiar words. The Bible reading was frequently raised at press conferences, and at one Borman recalled the Soviet cosmonaut who had said, 'I don't see God in the heavens.' 'I didn't see God either, but I saw evidence that God lives,' he responded. The evidence was that of science: 'the Earth looked at one time the way the Moon does now', and its transformation into something living provided direct evidence of the divine hand at work. Borman also compared 'the freedom of expression enjoyed by the Apollo 8 crew' with the allegedly prepared scripts read out by their Soviet counterparts.[46]

Press coverage of Apollo 8 enthusiastically picked up on the religious theme. 'For the first time large numbers of men are thinking of themselves as citizens of the Earth,' proclaimed the *Kansas City Star*. 'Perhaps we can turn this planet of ours into more of a spiritual as well as a physical oasis.' Recalling the Christmas season and the Genesis broadcast, the *Washington Post* mused that 'man has far to go here at home to fulfill the Christmas promise of Peace on Earth, Good Will toward Men.' 'Why did it move us so?' asked Max Lerner in the Washington *Evening Star*. His answer was that the poignant combination of the barren Moon, the lonely oasis of Earth and the simple biblical message was a reminder of 'man's arrogance in the face of mysteries he pretends to understand'. 'All of this is science?' asked the conservative commentator William F. Buckley. 'Don't believe it. We have reached into God's territory.'[47]

Not everyone agreed, however. Even as Apollo 8 returned to Earth, the atheist campaigner Madalyn Murray O'Hair announced that she would bring a lawsuit against NASA for 'evangelising' by broadcasting prayers from space, and shortly afterwards founded a national Center for Atheism to back her up. NASA was worried. In 1963 O'Hair had succeeded in getting the US Supreme Court to rule 8 to 1 that prayers in state schools were unconstitutional. When the Apollo 8 astronauts addressed Congress on 9 January Borman joked: 'One of the things that was truly historic was that we were able to get good Roman Catholic Bill Anders to read the first four verses of the King James Bible. But now that I see the gentlemen here in the front row' – the Supreme Court judges were there – 'I am not sure we should have read from the Bible at all.' A NASA official anxiously went through the transcript of communications, heavily annotating the Genesis broadcast. Houston had replied, 'Thank you for a very good show.' 'So CapCom did make a comment on how good the show was' read the marginal note.[48]

NASA took care to gather evidence that a separate prayer which Borman had read from lunar orbit on Christmas Eve had been a purely private arrangement. Borman had arranged to alert Rod Rose, the primary mission planner at Houston, with the codeword 'Experiment P-1' and Rose got ready to record the prayer, which was later played back to the congregation of the Episcopalian church which they both attended.[49] The Gideon Bible Society excitedly (but mistakenly) claimed that the astronauts had read from its own Bibles; NASA politely but firmly put it right. Borman received 34 letters of complaint about the Genesis broadcast, but nearly 100,000 letters in favour, many of them denouncing O'Hair and apparently coordinated, from all over the world. There were so many that NASA drew up a standard reply. Loretta Lee Frye of Detroit, a member of the Houston-based Apollo Prayer League, started a petition in support of Bible readings in space, which was presented to the head of NASA three months later with over half a million signatures. Four years on, the movement claimed 10 million supporters and had sent Bibles to several astronauts to be left on the Moon. All this began to resemble the evangelical agitation which had accompanied the Mercury missions, but this time the responsibility seems to have lain not so much with the astronauts as with the tactless Madalyn Murray O'Hair. To NASA's relief, the Court at length ruled that 'the first amendment does not require the state to be hostile to religion, but only neutral'. As long as the astronauts spoke in a personal capacity, the relevant principle was their own freedom of speech.[50]

The Genesis reading and the Earthrise picture so closely associated with it seemed to carry a universal message. On Christmas Day, as well as Archibald MacLeish's secular meditation 'Riders on the Earth', the *New York Times* ran a column by Edward Fiske arguing that a scientific age needed to heed the call of liberal theologians for 'a purified, primitive Christianity', purged of supernaturalism and welcoming to scientific discovery. In London, the Archbishop of Canterbury agreed that space travel 'should in fact strengthen our sense of dependence upon God as the creator of the universe'. The *Los Angeles Times* sensed that 'the flight of the astronauts produced great mental and spiritual ferment among men . . . turning their thoughts inward on their own condition and that of their troubled planet'. Religion had grown up and was no longer threatened by science.[51]

The Earthrise image got its biggest circulation of 1969 on a postage stamp – 'the stamp that everybody swooned over,' recalled Bill Anders. The idea was suggested by NASA historian Eugene Emme on 9 January, and was conveyed so rapidly up to the Vice-President, the President and the Postmaster-General that an announcement was made the next day. The choice of the Earthrise image was almost automatic, but an early draft of the stamp shows a blank background. Only after further discussion were the words 'In the beginning, God . . .' added. At the launch, the Postmaster-General made a self-important speech about bringing the US Postal Service into the space age. It says something for the impact of Apollo 8 that the most significant feature was hardly commented upon: it was the first time since the 1963 O'Hair ruling that any religious wording had appeared on a US postage stamp. Perhaps emboldened by the precedent, President Nixon wanted to alter the plaque left on the Moon by Apollo 11 to read 'We came in peace *under God* for all mankind', but NASA blocked him until it was too late to change it.[52]

While a fair amount of old-fashioned evangelical polemic surrounded Apollo 8, then, its main effect was to boost the trend towards a more universalist and science-friendly version of Christianity, a trend in which the Roman Catholic Church, then led by its most liberal pope, was in the vanguard. One detail of the mission symbolised this: the Intelsat-3A satellite, rushed into service to cover the Apollo 8 mission. It had gone live just in time to cover the Genesis broadcast, but its very first transmission, only a few hours before, was the Pope's Christmas Mass relayed live from Taranto.[53]

There was a poignant postscript to all this, which sheds some light on the cool relationship between the first two men on the Moon. Buzz Aldrin, impressed by the Apollo 8 prayer and Genesis reading, sought agreement from Deke Slayton to read his own choice of passage from the Bible ('What is man that thou art mindful of him?') as Apollo 11 made its way to the Moon, and to conduct and broadcast Holy Communion from the lunar surface. With the O'Hair legal action still pending, NASA refused. He should avoid all overt religious references, but he could read the prayer on the way back when it would be unlikely to arouse hostile notice. Aldrin told Armstrong of his plans; Armstrong offered no approval and raised no objection. At Tranquility Base on board the Eagle, in the quiet hours between touchdown and Moonwalk, Aldrin invited those listening 'to contemplate for a moment the events of the last few hours and to give thanks in his own individual way'. He then turned off his microphone, pulled down a folding shelf to use as an altar, and unwrapped the Communion kit provided by his Presbyterian minister back on Earth: a small silver chalice, a tiny vial of red wine and a wafer. 'I had plenty of things to keep busy with. I just let him do his own thing,' said Armstrong. One of the only two men on the Moon celebrated the first extraterrestrial Holy Communion alone, ending the brief ceremony with words from St John's gospel, 'apart from me you can do nothing'. It was not exactly an advertisement for the unity of mankind.[54]

Between them, all these three wider contexts – Earth science, the freedom of space and the Genesis broadcast – ensured that the Earthrise photo resonated powerfully across the planet. The sense of undogmatic spirituality conferred by the Christmas Eve reading from Genesis fitted nicely with the newly recognised peace of space, and with the astronauts' status as 'the envoys of mankind'. The rediscovery of Earth harked back to the origins of the space programme in the IGY project of a decade earlier. Commentators in 1968 wrote that, after the traumatic events of that year in the US, Apollo 8's pioneering voyage away from the Earth came as a great sigh of relief. But more deeply than that, with Apollo 8 humankind saw that it had escaped from the most dangerous phase of the nuclear arms race, entered the heavens and been rewarded with a God's eye view of Earth.

It is possible to imagine an astrofuturist version of the story in which the sight of the tiny, distant Earth receding over the lunar horizon was promoted as a kind of farewell postcard from the future: Earthset rather than Earthrise. Such a view certainly had its adherents. Even more people would have identified with a feel-good 'one world' message, as NASA's publicists had suggested. The crew of Apollo 8 inclined in that direction at first, but drew back in favour of a simple reading from the Book of Genesis. Earthrise therefore became associated with the Creation story, and with the feeling that the ultimate achievements of technology brought humanity face to face with ultimate truths. Confronted by the reality of the Earth from space, the fuzzy 'one world' idea was tried and found wanting. Its place was supplied by a sharper concept, more ancient than 'one world' and more obviously visual: the concept of the whole Earth.

CHAPTER 8

From Spaceship Earth to Mother Earth

At the beginning of the first space age Hannah Arendt posed a timely question: 'Has man's conquest of space increased or decreased his stature?' Arendt, who would later write searching studies of Nazi war criminals, was alarmed by those who welcomed space flight as an 'escape from men's imprisonment to the Earth'. True, philosophers had always dreamed of viewing mankind from some 'Archimedean point' outside the Earth, but, she mused, their speculations were inspired by a harmonious desire to 'discover the overall beauty and order of the whole', not by so destructive an impulse as the desire to 'conquer space'. Seen from orbit, people would lose first their individuality, then their humanity. Drivers would become indistinguishable from their cars, like so many metal-shelled snails. City-dwellers would appear like hordes of laboratory rats, scuttling in their mazes. People would become populations, subjects for statistical study rather than philosophy. Mankind's stature would be miserably reduced. Secularisation had already dethroned God the heavenly father; Mother Earth would be next. For Arendt, Earth was not a prison to be escaped but 'the very quintessence of the human condition . . . earthly nature, for all we know, may be unique in the universe'. Space travel was nothing less than a 'rebellion against human existence'.[1]

The conservative columnist William F. Buckley had a name for such humanist-inclined space-worriers: 'flat-Earth liberals'. There were plenty of them. 'After Sputnik,' commented the media guru Marshal McLuhan, 'there is no nature, only art . . . the Earth is an old nose cone.' 'The Moon is an escape from our Earthly responsibilities . . . it leaves a troubled conscience,' wrote *New York Times* columnist Anthony Lewis at the time of the first Moon landing. At the end of it all, he feared, 'we would know the stars but we would not know ourselves'. Kurt Vonnegut was deeply suspicious of the whole enterprise: 'Earth is such a pretty blue planet in the pictures NASA sent me. It looks so *clean*. You can't see all the hungry, angry earthlings down here – and the smoke and the sewage and trash and sophisticated weaponry.' The most hostile critic made the most prophetic comment. Denouncing the manned space programme as *The Moon-Doggle*, the sociologist Amitai Etzioni urged: 'As we move deeper into space, we should be facing the Earth.'[2]

Facing the Earth was exactly how the space programme ended up. The idea that the Earth was a finite system which needed to be carefully managed had been taking shape in the generation after the Second World War, in fields of thought as diverse as cybernetics, ecology and theology, but it only really took off when the sight of the whole Earth gave humanity a picture to think with. Hannah Arendt need not have worried so much. The space programme changed thinking about the Earth, but not in the way that either its supporters or its critics expected. The Apollo years of 1968–72 coincided almost exactly with the take-off of the environmental movement. Earthrise was followed by Earth Day. As men journeyed from the Earth to the Moon, the human race made the philosophical leap from Spaceship Earth to Mother Earth.

Spaceship Earth

Looking back, one of the most striking aspects of the first space age is the willingness of advanced thinkers to write off the Earth. This was also the time of greatest nuclear peril, of 'mutually assured destruction', the Cuban Missile Crisis and the film *Dr Strangelove*, when the futurologist Herman Kahn in his books *Thinking the Unthinkable* and *On Thermonuclear War* coined the term 'megadeaths' and took 'the shroud measurements of the corpse of civilisation'.[3] Nuclear missiles and space rockets were close cousins in an explosion of technology that threatened to belittle if not actually extinguish humankind. Kubrick's Doctor Strangelove had looked forward to preserving the core of a master race from nuclear holocaust in a nuclear bunker.

The idea that the peoples of Earth were bound together by common interest on a small and vulnerable planet originated not with the environmental movement but with fear of nuclear war. The head of Transworld Airlines, speaking at the first International Globe conference in 1962, noted that as the world shrank through air travel the threats to it were growing: 'We are all of us in the same boat, for better or for worse.'[4] When President Kennedy said in the same year that Apollo 'might hold the key to our future on Earth', he had in mind surviving the arms race. This was, after all, the year of the Cuban Missile Crisis. Colonies in space, like a high-tech version of Noah's Ark, were one way out of the nuclear holocaust.

Similar thoughts had recently been advanced in a 1961 book called *H. G. Wells and the World State* by W. W. Wagar. Wells in his 1913 novel *The World Set Free* had envisaged an atomic war, after which mankind would at last allow its best men to come forward and design a better world. There was more than a suspicion that Wells would have welcomed such a turn of events.[5] Wagar wrote:

> I suggest quite seriously that one fragment of the world revolutionary movement should detach itself from the main body at a very early stage and direct its energies towards the building of an ark of civilisation, a renewal colony well enough staffed and supplied to guide the survivors of a total war back to civilised life and forward to human unity . . . Homo Sapiens would become the new Neanderthals to the new scientific Cro-Magnons.

Wagar felt that 'the alienated, poor and oppressed' would somehow be the beneficiaries of this new 'organic world civilisation'. In his influential 1929 lecture 'The world, the flesh and the devil', the evolutionary Marxist thinker J. D. Bernal regretted that humankind inhabited 'a world limited in space to the surface of the globe'. He foresaw colonies of 30,000 people, orbiting the Earth in transparent goldfish bowls ten miles across. Life here might seem 'extremely dull', conceded Bernal, but civilisation would have evolved to match.[6]

The idealisation of space as a place to live went back to Konstantin Tsiolkovsky. Writing in the early twentieth century, he contrasted the overcrowded Earth, with its 'back-breaking and painful trouble', to the 'vast and free . . . space that surrounds our Earth . . . filled with light. . . . Who is there to stop men from building their greenhouses and their palaces here, and living in peace and plenty?' Tsiolkovsky was desperate to escape the land-bound existence of the Russian peasant, and although he worked out all the technological details this was still in a way a traditional peasant vision of heaven, but there were answering spirits in the affluent West. 'Looking out across immensity to the great suns and circling planets, to worlds of infinite mystery and promise,' the young Arthur C. Clarke wrote in the 1930s, 'can you believe that Man is to spend all his days cooped and crawling on the surface of this tiny Earth – this moist pebble with its clinging film of air? Or do you, on the other hand, believe that his destiny is indeed among the stars, and that one day our descendants will bridge the seas of space?'[7]

'If the Sun dies', worried the science fiction writer Ray Bradbury, human civilisation would die with it: 'Let us prepare ourselves to escape, to continue life and rebuild our cities on other planets: we shall not be long of this Earth! . . . Let us forget the Earth.' Bradbury's interviewer, Oriana Fallaci, was taken aback. Was the Earth a prison, then? 'All right, I'm quite comfortable in it, it is warm and safe, like a maternal womb. But you can't stay in your mother's womb forever . . . the only way the Earth can continue her life is by spitting you out, vomiting you up into the sky, beyond the atmosphere into worlds you cannot imagine.' He spoke, noted Fallaci, 'in a low voice, his eyes half shut . . . like a priest who recites the Pater Noster'.[8]

With the space programme taking off, such visions seemed only a step away from being realised. For Louis J. Halle in the *New Republic*, Apollo represented 'man's liberation from this earthly prison'. Human beings were 'like intelligent creatures confined to the ocean deeps . . . at last we are beginning to escape'. In the year of Apollo 11, the Princeton physicist Gerard O'Neill came to understand Earth as a gravity well, 'a hole which is 4,000 miles deep'. He went on in the early 1970s to devise 'space colonies', 'inside-out planets' capable of holding Earth's surplus population. 'We should ask, critically, with an appeal to numbers, whether the best site for a growing, advanced industrial society is Earth, the Moon, Mars, some other planet, or somewhere else entirely. Surprisingly, the answer will be inescapable: "somewhere else entirely".' The crucial step, felt O'Neill, was to lose 'the planetary hang-up'; 'goodbye Earth!' was the breezy slogan of his supporters. Among them was Isaac Asimov, who dismissed opposition to the scheme as 'planetary chauvinism'. Jesco von Puttkamer, an associate of Wernher von Braun who led NASA's Advanced Programs Office in the 1970s, liked to quote the words of the biologist and mystic Teilhard de Chardin: 'Man has no value save for that part of himself which passes into the universe.'[9] In the astro-futurist view, the sight of the distant Earth was proof that humankind had got itself in cosmic perspective at last and was, as it were, on its way out.

Technological ambitions for controlling the Earth were also at their peak during the first space age. There were plans to use nuclear explosives to mine minerals and dig a second Panama Canal. The Earth was to be wired, cabled and pipelined, its fertile lands transformed by a 'green revolution', its seas farmed and its human communities united in a single 'global village'. Perhaps this was why the first astronauts were so surprised to find so few marks of human activity from space. The catastrophic national rivalries of the first half of the century were widely believed to be giving way to an age of global planning and international government that would gradually bring the Earth under rational control. The new sciences of systems theory and cybernetics brought a confident, apolitical understanding of organisations, governments, populations and ecosystems. NASA was itself a monument to belief in the power of systems-based management – 'space-age management', as NASA's chief James Webb liked to call it.[10]

As the first satellite systems were beginning to connect up around the Earth, the media guru Marshall McLuhan wrote: 'Today we have extended our central nervous system itself in a global embrace, abolishing both space and time as far as our planet is concerned.' His vision of a shrinking Earth owed something to the manned space programme, which, he suggested, had 'altered man's relation to the planet, reducing its scope to the extent of an evening stroll'.[11] In 1966 Senator Joseph E. Karth, chair of the House Committee on Space Sciences, made a perceptive comment. 'One important contribution of aerospace technology has been general acceptance of the "total system approach",' he told the National Space Club. 'A closely connected idea is the growing awareness that we need to view the world environment as a whole. And I have a hunch that in retrospect, historians will consider this concept as one of the truly significant ideas of our century.'[12] Soon, it seemed, the entire globe could be managed like a single giant system. The system acquired a name: 'Spaceship Earth'.

'Who first realized that the Earth was a spaceship?' asked the ecologist Garret Hardin in 1972. His answer was simple: 'Nobody was first. Great visions grow slowly in the minds of men.' By then the term had become identified with the environmental movement, and it has often been assumed to have been a response to the Earthrise photograph. 'Spaceship Earth', however, did have an owner. Long before men left the Earth, Buckminster Fuller had thought his way out of it. Fuller was one of the great original thinkers of the twentieth century, and had coined the expression 'Spaceship Earth' (he claimed) as long ago as 1951. His most famous invention was the ultra-light geodesic dome, an icon of space-age architecture but inspired by nature. Fuller's geodesic domes could and did house just about anything, from corporate headquarters and international expos to communes and happenings. Described as a poet of technology and as everyone's 'crazy-inventor grandfather', he sought to understand the world as a system, nature and humanity together. He saw himself as standing outside it in the role of 'comprehensive designer', anticipating and shaping human progress.[13]

When Walter Cronkite, covering Apollo 8, described the Earth as 'floating in space', Fuller asked scornfully: 'floating in what?' The Earth moved through space just like everything else: 'All of us *are*, always *have been*, and so long as we exist, *always will be – nothing else but – astronauts*.' (Although an admirer of economy in nature, Fuller was also a windbag.) He derided the 'geometrically illiterate' folk, from the President downwards, who could look at the Apollo 8 photographs and still talk about going 'up' to the Moon and 'down to Earth': 'in which direction of Universe is DOWN located?' In the same spirit, the photographic historian Beaumont Newhall wrote that every photograph taken through a telescope 'is indeed a space photograph – from spacecraft Earth'. When people asked where he lived, Fuller would reply: 'I live on a little planet called Earth. I never leave home.'[14] If antiquated ideas of Earth persisted, proclaimed Fuller, 'humanity is doomed. But there is hope in sight. The young! . . . There is a good possibility that they may take over and successfully operate SPACESHIP EARTH.'

That was exactly what started to happen. The view of Earth from Apollo 8 revived interest in Buckminster Fuller's 'World Game', a computer model designed to work out 'an efficient process for managing Spaceship Earth'. Fuller had first proposed the idea five years before, to be played by button-pushing visitors on a world map wired with light bulbs at Expo '67; the organisers accepted his geodesic dome but rejected the game. In the summer of 1969 the 'World Game' was played by teams of idealistic students at universities and colleges across the US, Canada and Europe. *Mother Earth News* described it as 'a unique experiment to develop a computer coordinated model of planet Earth – complete with resources, history, human attitudes and social trends – that can be used to "play the world" and develop ways of running the future for the benefit of all mankind'. One player, a former Dominican brother, explained:

We start out from the basic premise that all the problems of today's world are in some ways brought about by compartmentalized thinking, that boundaries and territories create great barriers between people and that most of the waste of resources is brought about by adherence to these political myths, these political boundaries. We're going to have to think of the Earth as one island, one nation, one boundary.[15]

Fuller's vision of Earth was a technocratic one (one senses that pets were as absent from his Spaceship Earth as they were from his geodesic domes), but well before anyone saw the whole Earth he was ready with the ideas to think about it. One of his most enthusiastic followers was Stewart Brand, founder of the *Whole Earth Catalog*. Fuller's vision of 'Spaceship Earth' was brought to life by the first picture of the whole Earth, but it was Brand who popularised it as an icon for the eco-aware counterculture that mushroomed in the Apollo period.

The *Whole Earth Catalog*

Stewart Brand's remarkable *Whole Earth Catalog* (1968–71) was a magazine-format directory offering 'access to tools' for the counterculture that was rapidly spreading from the communes and comics of California to the rest of the world. From the cover stared an image of the whole Earth from space. The Apollo 8 Earthrise, it was explained inside, 'established our planetary facthood and beauty and rareness (dry Moon, barren space) and began to bend human consciousness'. Under the picture of Earth on the back cover of later editions was the slogan, 'We can't put it together. It is together.' In the mid-1960s Brand was one of a group of artists staging 'happenings' on the American West Coast, hanging out with Ken Kesey's 'Merry Pranksters' and helping set up the legendary 'Electric Kool-Aid Acid Tests'. 'I'm a former ecology student,' he told a Congressional committee in 1970, 'and I can report that ecology as a science is pretty boring.... Ecology as a movement, as a religion, is tremendously exciting, and everyone can get a piece of the fervor.' He wanted to see environmental spending directed towards 'the space program which has given us the anti-environmental perspective to see our planet whole and alive and in hazard'.[16] The *Whole Earth Catalog* acted as a link between techno-utopianism, environmentalism and the counterculture, brought together under the image of the planet. It came much closer than Buckminster Fuller's own treatise to being the 'Operating manual for Spaceship Earth.'

The first *Whole Earth Catalog* appeared in autumn 1968, in an edition of only a thousand copies, but it became a publishing phenomenon. After several editions and supplements, *The Last Whole Earth Catalog* was published internationally by Penguin in 1971 and sold nearly a million copies. Steve Jobs, the founder of Apple computers, described it as 'like Google in paperback form ... idealistic, and overflowing with neat tools and great notions'.[17] Inside the covers was everything for the alternative lifestyle, ranging from teepees and water purifiers ('recycle your piss') to ornaments and manuals. The printed exhibits included Buckminster Fuller's *Operating Manual for Spaceship Earth*, a poster of the Andromeda Galaxy, Olaf Stapledon's cosmic novel *Star Maker* ('a true vision'), Loren Eiseley's philosophical work *The Unexpected Universe*, an *Atlas of the Universe*, several books of photographs from space, including *Ecological Surveys from Space* and NASA's *This Island Earth*, a manual on *The Biosphere* and a book of nude photos tastefully entitled *Surface Anatomy*. There was even a summary of the provisions of the 1967 Outer Space Treaty. Posters of the whole Earth were offered for sale at a discount of 50 per cent for five or more.[18]

Brand had had the idea for the *Whole Earth Catalog* in the spring of 1966, while reading Barbara Ward's newly published environmentalist manifesto *Spaceship Earth*.[19] Ward, like Brand, had been influenced by Buckminster Fuller, so something was likely to click. The *Whole Earth Catalog* when it appeared two years later began with a section on 'understanding whole systems' which promoted the works of Buckminster Fuller and other big ideas merchants. Systems thinkers, space enthusiasts and New Age mystics alike could all welcome the sudden, transforming impact of the view of the whole Earth as the ultimate example of how the experience of a few individuals could be disseminated by technology throughout the whole 'global village'.

But what image of the Earth did the *Whole Earth Catalog* actually use, and where did it come from? Later editions used the Apollo 8 Earthrise, and when this had become familiar the last one went for the Apollo 4 photograph of the crescent Earth. But Earthrise had not been taken when the first edition was published, making the image on its cover all the more novel: a colour reproduction of the 1967 ATS-III satellite TV picture of Earth. Some years later Brand told a conference of the Lindisfarne Association the story of how (he believed) he had played a behind-the-scenes role in causing that same picture to be taken.

> I was sitting on a gravelly roof in San Francisco's North Beach. It was February 1966. Ken Kesey and the Merry Pranksters were waning toward Mexico. I was twenty-eight.
>
> In those days, the standard response to boredom and uncertainty was LSD followed by grandiose scheming. So there I sat, wrapped in a blanket in the chill afternoon Sun, trembling with cold and inchoate emotion, gazing at the San Francisco Skyline, waiting for my vision.
>
> The buildings were not parallel – because the Earth curved under them, and me, and all of us; it closed on itself. I remembered that Buckminster Fuller had been harping on this at a recent lecture – that people perceived the Earth as flat and infinite, and that that was the root of all their misbehavior. Now from

my altitude of three stories and one hundred miles, I could see that it was curved, think it, and finally feel it.

But how to broadcast it?

Brand began a campaign to pose the question: 'Why haven't we seen a photograph of the whole Earth yet?' Although a fan of the space programme, he felt something was wrong. 'There we were in 1966, having seen a lot of the Moon and a lot of hunks of the Earth, but never the complete mandala . . . and it was a bit odd that for ten years, with all the photographic apparatus in the world, we hadn't turned the cameras that 180 degrees to look back.' Brand had his question printed on button badges, and on a poster which showed a star field with a hole cut in the middle, playing on the American instinct for conspiracy. He posted the badges to Congressmen, officials at the UN and NASA, Soviet scientists and diplomats, and to Buckminster Fuller and Marshall McLuhan.[20] Then he turned up at the University of California's Berkeley campus (reported the *San Francisco Chronicle*) 'wearing his shocking pink and blue sandwich boards, white coveralls, desert boots and a top hat decorated with a yellow flower . . . he gathered a sizeable number of students around him. He also sold quite a few lapel pins for 25 cents apiece.'[21] He was thrown out by James L. Sicheneder, 'Dean Fuzz', the campus policeman appointed to keep political activists off the university campus. Brand nonetheless returned several times before moving on to Stanford, Columbia, Harvard and MIT (where his brother was an instructor), conducting what he called 'street-clown seminars on space and civilization'.

Brand was in the right place at the right time, for California (including its university) was one of the hubs of space technology, and its population was very space aware. The *San Francisco Chronicle* of this period was full of stories about space missions, UFOs, planetarium shows (including 'the death of the Earth') and the weather pictures transmitted to San Francisco by one of the new weather satellites. At the same time, suspicion of government was fed by the long-running story of the Palomares incident, where the US Air Force managed to lose two hydrogen bombs in an aircraft crash in Spain and vainly tried to cover up what had happened. If Stewart Brand's question was going to resonate anywhere, it was in California. But did the vibes reach NASA?

For his mass-mailing of badges, Brand looked up the names and addresses of 'all the relevant NASA officials'. At Berkeley the astrophysicist George Field bought five badges to take to NASA, Philip Morrison (who also had NASA connections) bought two in Harvard Square,[22] and at Stanford Brand met people from the NASA Ames Research Center. Brand claimed later to have met the military intelligence officer detailed by Washington to investigate him; California, the spook had reported, was full of harmless individualists. Some NASA people must have known of his campaign. Alas, no recollections from the NASA side have so far been recorded. The story of how the photograph came to be taken does not require outside input and (as we saw in chapter 4) senior NASA officials had asked Richard Underwood to 'get great photographs of Earth' after the Gemini spacewalk photographs a year earlier. But Underwood did not work on the Lunar Orbiter project, and it seems a reasonable guess that Brand's campaign helped open more minds at NASA to the idea of an Earth photograph not long before it became technically feasible. Whether by coincidence or not, a few months after Brand's campaign NASA's Lunar Orbiter spacecraft turned its cameras back towards Earth.

Perhaps the *Whole Earth Catalog*'s most unusual exhibit was a full-colour film called *Full Earth* taken from NASA's geostationary ATS-III satellite, 23,000 miles up. It was described as a 'home movie' of the Earth over twenty-four hours: 'You see darkness, then a crescent of dawn, then advancing daylight and immense weather patterns whorling and creeping on the spherical surface, then the full round mandala Earth of noon, then gibbous afternoon, crescent twilight, and darkness again.' There was then a series of close-ups of different regions of the Earth, blurring and swirling psychedelically like the oil lens projections which were then lighting up underground concerts by the likes of Pink Floyd. Like the LSD that helped bring it to life, the image of the whole Earth started out as a project of the US government and ended up as a badge of the counterculture.[23]

The ATS-III satellite whole Earth on the front of the *Whole Earth Catalog* was, then, not just an icon but indirectly a trophy of Brand's campaigning skill. It exemplified the idea that in an age of technology, individual consciousness has the potential to change our view of the world. It was indeed an impressive demonstration.

Earth Day

As the Apollo astronauts began to return to Earth, the environmental movement took off. Earthrise was followed by Earth Day. Founded in 1970, revived in 1990, and still marked on calendars today, Earth Day has become a fixture of American life, and it can be traced directly back to the impact of Apollo 8's view of the Earth. Early environmentalism, while suspicious of big technology generally, was not particularly hostile to the space programme as such.[24] Earth Day filled a lull in the Apollo programme itself, for there were no successful missions to the Moon in 1970. Ironically, while there may have been only one Earth, there were two Earth Days.

The first Earth Day to get off the ground was organised by John McConnell, a longstanding peace activist and advocate of international co-operation in space. He had toyed with the idea of an 'Earth Day' as early as 1967 but nothing concrete came of it until the first photo of Earth appeared in *Life* in 1969. Then, he recounted, 'I . . . experienced in a deep and emotional way a new awareness of our planet . . . it occurred to me that an Earth flag could symbolize and encourage our new world view and that the Earth as seen from space was the best possible symbol for this purpose.' He called NASA's head of public affairs, who said, 'What a wonderful idea!' and sent him a transparency of the same photograph. McConnell copyrighted a version of the image showing Earth against a dark blue background, and later registered it as a trade mark. Five hundred flags the size of tea towels were ordered and sold, mostly among the crowds who gathered in New York's Central Park on 20 July 1969 to watch the first Moon landing.[25]

McConnell then went to California to organise the first Earth Day. It was proclaimed by the city and county of San Francisco on 21 March 1970, the day of the spring equinox when day and night were equal all over the world. It was to be 'a day to celebrate planet Earth: one home for mankind, one source of life, one responsibility for all. A day to begin our renewal of Earth life.' Supporters were urged to plant trees and flowers, clean up their neighbourhoods, observe a silent 'Earth Hour' of prayer or meditation at 11 a.m. Pacific Standard Time, go out and enjoy nature, and display the Earth flag as a reminder that 'each person has a basic right to use the

Earth, and an equal responsibility to build the Earth'. A button badge was issued showing the Earth with the slogan: 'Our inheritance – our responsibility', while another bore the slogan 'Give Earth a chance' (inspired by John Lennon's recent song, 'Give peace a chance').[26] Thanks to a mention in the *Whole Earth Catalog* there was a rush of demand for flags, and volunteers distributed over 15,000 of them, which were flown by schools, churches, ecology groups, businesses and youth organisations.

McConnell went on to found an Earth Society which continued its campaign for some years, devising idealistic schemes for global citizenship and world passports, and promoting the Earth flag as 'a non-government flag for all Earth people . . . to encourage equilibrium in nature, in social systems, and in the minds of men'. Governments were invited to place the Earth in the corner of their national flags and (recalled McConnell) 'the QE2 flew the Earth Flag in 1973 on a two-week Earth Society cruise in the Caribbean with Isaac Asimov, Carl Sagan, Burl Ives, and other Earth patriots'. The same year, with the 'Blue marble' still fresh in the public memory an Earth Day essay by the anthropologist Margaret Mead was syndicated internationally. On another occasion thirty large Earth flags were ordered for an event in Central Park to celebrate New York's ethnic diversity. An official confided later: 'It solved a problem. All you'd have to do is to leave out one ethnic flag and you'd have a crisis. This covers everybody.' Earth Day reached its twentieth anniversary in 1990, issuing the disconcerting image of a skull in red, white and blue with a full-colour Earth sitting in the brain pan: 'a peaceful place or so it looks from space' ran the caption. McConnell prepared an 'Earth People Proclamation' . . . by the people of Earth, for the people of Earth'. It warned a crisis of ecological balance and urged signatories to pledge 10 per cent of their income to campaigns for the environment and social justice – 1 per cent more than the proportion said to go on war and pollution.[27]

While the first Earth Day was idealistic and internationalist, the second was more practical and localist, and this one achieved the wider appeal. It was organised by the Democratic Senator for Wisconsin, Gaylord Nelson. In September 1963, the same month in which President John F. Kennedy addressed the United Nations on the need for international cooperation in space, he also joined Nelson on a national conservation tour, visiting eleven states in five days. The tour was Nelson's idea. Large parts of his home state of Wisconsin had been ruined by logging, and in 1961 Nelson had set up the Wisconsin Outdoor Recreation Program to reclaim land for environmentally sensitive tourism. The tour didn't have anything like the profile of Kennedy's pronouncements on space, but it helped prepare opinion for the landmark Water Quality Act of 1965 which began to accustom states to federal intervention on environmental issues.[28]

In the summer of 1969 – the summer of Apollo – Nelson was impressed by the wave of 'teach-ins' organised by opponents of the Vietnam War and on 20 September Nelson called for a wave of environmental teach-ins, enlisting (as it were) the services of Sam Brown, a 25-year-old anti-Vietnam activist. 'The response was electric,' Nelson recalled, 'it took off', pretty much organising itself from the senator's own office. Larry Rockefeller launched the fund with a donation of $1,000; not to be outdone, the presidents of United Auto Workers and the American Federation of Labor both called and gave $2,000 each. Other politicians hastened to acquire environmental credentials, all 'keeping up with Gaylord', as the *Washington Post* put it. Eventually $185,000 was raised. The consumer champion Ralph Nader lent his support, and when in January 1970 a Washington office opened it was provided by John Gardner, shortly to found the citizens' lobby organisation Common Cause. Another leading organiser was Denis Hayes, who went on to found Environmental Action. Earth Day became a nursery for green activists.[29]

Earth Day was held on 22 April 1970 as a mass environmental teach-in. The lecturers included Barry Commoner, Paul Ehrlich, René Dubos, Ralph Nader, Benjamin Spock and Allen Ginsberg. The organisers claimed that 2,000 colleges, 10,000 schools and 2,000 citizen groups had resolved to participate. Ten thousand people rallied at the Washington monument, and there were rallies of up to 25,000 in many major cities. In New York, Governor Nelson Rockefeller signed a bill to create a Department of Environmental Conservation, then addressed a rally at Prospect Park, Brooklyn, and rode away on a bike. Inside the headquarters of the giant company Con Edison 'everybody was wearing an Earth Day badge'. On Fifth Avenue three children were photographed holding a banner with a map of the Earth and the slogan 'love it or leave it', nicely adapting the redneck motto 'America, love it or leave it' to the space age. President Nixon, suggested that the environment could become 'the major concern of the American people in the decade of the '70s', and in his State of the Union address urged the American people to 'make our peace with nature'. The Environmental Protection Agency was founded that year, and the Clean Air Act and a raft of other environmental legislation soon followed.

On the whole, Earth Day was more celebration than protest. The organisers promoted credit card burning, modelled on the draft card burning of the anti-war movement, but some commentators felt it lacked the same passion. Earth Day was criticised by socialists and some others as a non-confrontational feel-good movement which diverted attention from other kinds of protest – a somewhat ignoble complaint which persisted for many years until environmentalism itself began setting the political agenda. Nonetheless, the Earth was a relatively safe issue in a period disfigured by conflicts over race and Vietnam. 'If the environment had any enemies they did not make themselves known,' said the *New York Times*. 'Suddenly we are all conservationists,' observed *Life*.[30] Twenty-five years later Gaylord Nelson was awarded the presidential medal of freedom. The citation recorded that 'as the father of Earth Day . . . he inspired us to remember that the stewardship of our natural resources is the stewardship of the American Dream'.

While Earth Day seemed suddenly to mushroom, the soil had been under preparation for several years. The environmentalist breakthrough work of 1962, Rachel Carson's *Silent Spring*, had exposed the effects of pesticides and other chemicals on wildlife and had derived much of its impact from a 'man versus nature' theme. At the same time, the new discipline of ecology was emerging. Ecology, however, was about more than just nature conservation. Its concern was with life as a set of interlocking systems, human society included – at once the greatest threat to the environment and the source of the solution. Like whole Earth utopianism, ecology owed much to cybernetics and systems science, and there developed a specific branch of 'systems ecology' through the work of the brothers Eugene and Howard Odum. Systems ecology focused on the concept of the 'ecosystem' as the basic unit of understanding, and was challenged by orthodox evolutionists, such as Richard

Dawkins, for whom the individual gene was the unit. The discipline split, but while the evolutionary ecologists were generally judged to have won the battle of ideas, the systems ecologists achieved a high profile and were ready and waiting when the sight of the Earth gave a new boost to holistic thought about the environment.

Buckminster Fuller's idea of 'Spaceship Earth' was first popularised not by Fuller himself but as the title of a book by the environmental philosopher Barbara Ward. She saw the human race as the crew of a single spaceship, and urged,

> We have to live as human beings if we are to survive the vast space voyage upon which we have been engaged for hundreds of millennia – but without noticing our condition. This space voyage is totally precarious. We depend upon a little envelope of soil and a rather larger envelope of atmosphere for life itself. And both can be contaminated and destroyed. . . . Rational behaviour is the condition of survival.

The 'Spaceship Earth' concept also proved popular at the United Nations. It had been used as early as 1965 by Adlai Stevenson, the American Ambassador to the UN, in urging the need for human unity. In September 1968, Unesco held an international conference on the 'Use and Conservation of the Biosphere', published in 1970. The UN Secretary-General, U Thant, warned, 'Spaceship Earth is left without a central guidance and stewardship.' The drama of Apollo 13 unfolded a week before the second Earth Day. As the stricken craft drifted through space, struggling to maintain its life support systems, the systems ecologist Eugene Odum felt it was a timely reminder of the plight of 'Spaceship Earth', illustrating the urgent need to understand the Earth in order to save it.[31]

The UN followed Earth Day by staging in Stockholm in June 1972 a Conference on the Human Environment – the first Earth Summit. Ward was commissioned to write the report, together with the microbiologist and philosopher René Dubos. It was later published as *Only One Earth*, with a NASA photograph of Earth on the cover. The project was described as 'a unique experiment in international collaboration', bringing together the work of over 150 consultants across the world. Opinions had varied enormously between experts, with nuclear energy being particularly divisive, but the overall message was of 'the essential unity and interdependence of both technosphere and biosphere'. 'As we enter the global phase of human evolution,' wrote the authors, 'it becomes obvious that each man has two countries, his own and the planet Earth.' When the returning Apollo 8 astronauts were interviewed about their experience on CBS television, René Dubos had been among those watching. 'What turned out was that their deepest emotion had been to see the Earth from space,' noticed Dubos. 'The phrase "theology of the Earth" thus came to me from the Apollo 8 astronauts' accounts of what they had seen from their space capsule.' There were, sensed Dubos, 'sacred relationships that link mankind to all the physical and living attributes of the Earth', an insight, he felt, which was 'worth the many billions of dollars spent on the manned space programme'.[32] Within three years of the first sight of the Earth, then, the insights which it generated had become the official stance of the United Nations.

The years 1968–72 also saw a more general eco-renaissance. There was, of course, nothing new about concern for the natural environment. Ever since the days when Henry Thoreau wrote about Walden Pond, and arguably since the time the first American settlers thanked God for the bounty of nature, regard for the environment had been part of the American ideal, if not always of the American way. The concern for much of the century was overwhelmingly with wildlife and open spaces, through organisations such as the Sierra Club (1895), the Wilderness Society (1935) and the World Wildlife Fund (1961). National environmental organisations were founded at a rate of a little over one a decade from the 1890s. The environmental alarm call sounded by publication of Rachel Carson's *Silent Spring* (1962) was followed by the foundation of the Environmental Defense Fund (1967) to support legal actions. As soon as the Earth became visible, however, it began to acquire friends, starting in 1969 with Friends of the Earth. The years 1969–72 saw no fewer than seven major national environmental organisations come into being. Activist and radical, they aimed at social and cultural change rather than simple preservation. They were also increasingly professional: there were only two full-time environmental lobbyists in Washington DC in 1969, but within a few years there were dozens.[33]

The focus of the new environmental movement was not 'wilderness' or 'nature' but 'the environment', with humankind very much in the picture. Thanks to Earth Day the word 'ecology', all but unknown in the press much before, filled eighty-six columns of the *New York Times* index for 1970. In Britain, the idea of ecology reached public consciousness in the autumn of 1969 through the BBC's influential Reith Lectures, given that year on the theme of 'Wilderness and Plenty' by the ecologist Frank Fraser Darling. 'Ecology caught on like a new religion among the young on college campuses across the country,' wrote one commentator. 1970 was designated European Conservation Year, and a rash of health food shops broke out.[34] Other influential publications of these years included Max Nicholson's *Environmental Revolution* (1969), Barry Commoner's *The Closing Circle* (1971), the culmination of years of warnings by 'the Paul Revere of ecology', and Garret Hardin's *Exploring New Ethics for Survival: The Voyage of the Spaceship Beagle* (1972). This included Hardin's influential 1968 essay 'The tragedy of the commons', which offered an analogy between Spaceship Earth and a lifeboat to argue that some short-term loss of liberty was justifiable to rescue the environment.

The realisation that the Earth and its resources were finite had a parallel impact in the world of economics. Kenneth Boulding, a British economist who was also a pioneer systems theorist, delivered a paper as early as 1966 on 'The economics of the coming Spaceship Earth'. He contrasted the 'cowboy economy' of uncontrolled production and consumption with 'the spaceman economy, in which the Earth has become a single spaceship, without unlimited reservoirs of anything . . . a cyclical ecological system', to be conserved rather than exploited.[35] In 1972 *The Limits to Growth* was published, a report on 'the predicament of mankind' produced by an international team of scientists, educators, economists, humanists, industrialists and civil servants which had first gathered in Paris in the spring of 1968. There was no explicit mention of the sight of the Earth, but one ambitious flow diagram attempted to model the 'whole world system' in over a hundred bubbles and almost as many arrows. The conclusion was that the Earth, thanks mainly to over-population, would reach the limits of economic growth within a century, and that the need for sustainable development was urgent. A second report two years later compared the global economy to a living organism, such as an oak tree, whose growth rate needed to slow down as it reached maturity; exponential growth had to be replaced by 'organic growth'. It passed on a blunt warning: 'The world has cancer and the cancer is man.'[36]

There was also an eclectic publication called *The Pulse of the Planet*, which described itself as 'the Smithsonian Institution's "state-of-the-planet" report to the global community'. It was inspired by the Apollo 8 view of the Earth, which (wrote the editors) 'dramatically showed that the Earth, too, is a spacecraft, adrift in the universe. Completely dependent on its own self-contained resources, our planet rides as precariously along the edge of disaster as any satellite launched by man. Ironically, then, at the very moment man gained the Moon, his thoughts turned back toward Earth.' The book offered a curious *Fortean Times*-style miscellany of 'the exciting, unusual, and often oddball events' of the previous four years – natural disasters, man-made environmental disasters, and ecological events – aiming to give a sense of 'the planet as an ecosystem'.[37]

The famous Apollo 17 'Blue marble' photograph appeared in December 1972, just in time to supply the environmental movement with its most powerful icon. It was, however, the Apollo 8 image of December 1968 that had started it all off. Both images owed much of their instant power to the way they tapped into a ready-made agenda: in the case of the 'Blue marble' it was the eco-renaissance; in the case of Earthrise it was 'Spaceship Earth.' What happened over the years in between was that natural metaphors for the planet began to take over from technological ones. Earth had risen indeed.

The return of Mother Earth

Among those watching the launch of the final Apollo mission in December 1972 were the cosmologist Carl Sagan and the pioneer of the New Age movement William Irwin Thompson. For both men it was a spiritual experience. As they sat together at the edge of a Florida lagoon, Sagan speculated about life on Mars and alternative universes, while attempting to communicate in beeps with a nearby dolphin. As the rocket was launched, Thompson felt 'the sheer joy of knowing that men were turning the tables on the heavens and rising that comet out of Earth'; he compared it to the elevation of the Host at the Mass. 'The faces of the scientists were the faces of men who had witnessed the Transfiguration on the Mount,' he noticed. With the 'Blue marble' photograph twenty-four hours from being taken, Thompson sensed that the event marked a 'new stage in human culture ... a planetary society' characterised by a combination of immense knowledge and unlimited spiritual potential.[38]

A few months later Thompson founded the Lindisfarne Association at Long Island, New York. Named after the remote Northumbrian monastery, 'a spiritual elite holding on to knowledge in a dark age', Lindisfarne aimed 'to effect an evolutionary transformation of human culture'. 'We are the climactic generation of human cultural evolution,' wrote Thompson in his book *Passages about Earth*, 'and in the microcosm of our lives the macrocosm of the evolution of the human race is playing itself out.' There was a race against time going on: would the new planetary consciousness set in before human damage to the environment became irreversible?[39] In 1974 the Lindisfarne Association mounted a conference on 'Planetary Culture', one of the founding events of the New Age movement. Among other things, the conference brought together the astronaut Russell Schweickart and the activist Stewart Brand in a remarkable joint session.

Brand had already wound up the *Whole Earth Catalog* believing that it had run its course, but changed his mind in 1974 and published an update which led to a string of further editions. In the same year he founded the ideas magazine *CoEvolution Quarterly* (later *Whole Earth Review*) which acted as a bridge between science and the counterculture. He was now influenced by another leading systems thinker, Gregory Bateson, who had been developing a view of the natural world as a series of interconnecting information systems. Bateson saw the individual as 'a servosystem coupled with its environment', and wrote of an 'ecology of mind' that spanned the planet. The insights of advanced individuals, he argued, could be spread to the whole species, and he called for urgent action to bring humanity and the natural world back into balance. This emphasis on the power of individual consciousness to change society appealed to Brand, and indeed to a whole generation of countercultural radicals, environmentalists and 'New Agers' looking for new models of social change. Echoes of the 'ecology of mind' concept can be detected in the environmentalist slogan 'think globally, act locally', and in Russell Schweickart's conviction that on his spacewalk he was 'the sensing organ for mankind'. While he continued to take an interest in Buckminster Fuller's World Game, Brand was moving beyond the 'Spaceship Earth' concept and now favoured the idea of the Earth as Gaia, a living organism. 'The paradigm shift is from an engineering metaphor to a biological metaphor,' he declared in 1976.[40]

Russell Schweickart (as we saw in chapter 7) had been deeply affected by the sight of Earth on his spacewalk on Apollo 9. He had gone on to work as back-up to the Apollo-Skylab programme, which had recently ended with the crew successfully demanding more time to look out of the window at Earth. He had for some time felt an affinity with the disaffected youth movement, 'the only real hope for the planet', and had grown his hair. He was now Director of User Affairs in the Office of Applications at NASA's Washington DC headquarters, responsible for mediating between NASA and the users of space technology. Among other things, he had brokered the collaboration between Jacques Cousteau and NASA's Landsat project. When he also introduced Stewart Brand to Cousteau and NASA officials in Washington, the results of the encounter duly reached the pages of *CoEvolution Quarterly*.[41]

Brand told the Lindisfarne conference about his campaign to persuade NASA to take a picture of the whole Earth; his suggestion that the counterculture was ultimately responsible for the image which it had adopted as its own went down well. Nothing, however, could have prepared the audience for Russell Schweickart's remarkable meditation 'No frames, no boundaries'. He conveyed the sheer spiritual force of the experience of floating outside a spacecraft and watching the Earth revolve beneath him, Brand reported, in a stream of words 'like a long, pauseless prayer . . . Schweickart himself seemed amazed at what he was saying, amazed at the gathering he was attending, amazed – still – at the events which led him to drift bodily free between Earth and Universe. Remember the Star Child at the end of *2001*? Like that.' He closed with a few lines by e. e. cummings which he said had 'become a part of me somehow', thanking God for 'a blue true dream of sky; and for everything which is natural which is infinite which is yes'. Brand published the speech in *CoEvolution Quarterly* and the Lindisfarne Association sold tapes of it; it became one of the most famous of astronauts' testimonies. Schweickart later summed up his sense of the Earth in one short phrase: 'It's home, it's whole, it's holy.'[42]

In the mid-1970s, Brand's techno-friendly part of the counter-culture split over the issue of space colonies. Gerard O'Neill had published the first draft of his ambitious scheme in 1974, and then approached the Portola Institute in San Francisco, publisher of the *Whole Earth Catalog*, for support. Portola laid on a conference and Brand's Point Foundation provided a grant, funded from the proceeds of the *Whole Earth Catalog*: the counterculture was now helping to fund the space programme. Grounding his argument in contemporary concerns about overpopulation, energy shortage and the limits to growth, O'Neill argued that society could only continue to grow and therefore to serve its poor by moving into space to mine the Moon and asteroids and harness the unlimited energy of the Sun. He proposed shifting populations to 'inside-out planets', cylindrical colonies gravitationally suspended between Earth and Moon. Starting at 10,000 people, they could in time provide a home to millions. 'The subject is FREE SPACE', proclaimed Brand. 'For those who long for the harshest freedoms, or who believe with Buckminster Fuller that a culture's creativity requires an Outlaw Area, Free Space becomes what the oceans have ceased to be – [an] Outlaw Area too big and dilute for national control.' One illustration showed a curving, hilly landscape remarkably similar to the San Francisco Bay area, complete with estuary, suspension bridge and picnicking hippies.

The space colonies scheme provoked a fierce debate, splitting the readership of *CoEvolution Quarterly*.[43] There were some interesting supporters for the 'new planet' scheme. The microbiologist Lynn Margulis, an early Gaian thinker, was mildly in favour because of the invigorating effect of colonies generally: 'The values of the Mother Country must take on a new perspective from the distance of the colony. How small, idiosyncratic, isolating, anti-rational, parochial and repressive seem all tribal and national and imperial customs seen from a distance.' The cosmologist Carl Sagan preferred the term 'space cities' to colonies ('no Space natives are being colonized', emphasised Brand) and offered a countercultural version of the American frontier hypothesis.

The Earth is almost fully explored and culturally homogenized. There are few places to which the discontented cutting edge of mankind can emigrate. . . . Space cities provide a kind of America in the skies, an opportunity for affinity groups to develop alternative cultural, social, political, economic and technical lifestyles . . . . But this goal requires an early commitment to the encouragement of cultural diversity.

Russell Schweickart was another enthusiast. Pictures of Earth alone in space were one thing, but 'with O'Neill's "seedpods" . . . mother Earth need no longer remain barren'.

But while many readers looked forward to communes in space, many more were taken aback. 'I'd like to get away from Earth awhile/ And then come back to it and begin over,' wrote the poet Robert Frost. 'Earth's the right place for love: I don't know where it's likely to go better.' 'I like this planet,' wrote Richard Brautigan. 'It's my home and I think it needs our attention and our love. Let the stars wait a little longer. They are good at it.' One correspondent hit the nail on the head: 'Do you really think the government is going to let a funky rabble on their new space colony?'[44] NASA studied the scheme for it suggested another possible use for its main post-Apollo project, the reusable space shuttle. Despite Brand's advocacy, the environmentalists of the later 1970s mostly turned their backs on space. 'Small is beautiful' was the new idea; big technology was out, ecology was in. In California, however, the attempt to find a synthesis continued.

If 1970 was the year of Earth Day, 1977, in California at least, was the year of Space Day. Its theme was 'Ecology and technology find a unity in space'. By then the hippy-hating Ronald Reagan had been succeeded as state governor by Jerry Brown, a liberal figure who sought to build bridges with environmentalism and the counterculture. Brown was also the first major political figure after John F. Kennedy to offer a vision of space adventure. Having come under fire from both the environmental lobby, for approving a dam project, and from the business lobby, for allegedly driving Dow Chemicals out of the state by overregulation, Brown tried to bring them both together. Space Day was sponsored by California's aerospace industry, worried about threatened cuts to NASA's budget. Among Brown's advisers was Stewart Brand, who in turn persuaded the governor to hire Jacques Cousteau and Russell Schweickart; Schweickart spent the next two years as the governor's Assistant for Science and Technology.

California's Space Day took place at the Los Angeles Museum of Science and Industry on 11 August 1977, before a large invited audience. The speakers included the new head of NASA, Robert Frosch, Gerard O'Neill, Carl Sagan, Jacques Cousteau, the head of the Jet Propulsion Laboratory Bruce Murray (a fan of E. F. Shumacher and Ivan Illich), and Robert Anderson, the head of the company that was building the space shuttle. Stewart Brand chaired the debates. Brown, echoing Carl Sagan, offered a liberal take on the old astro-futurist theme of the space frontier. 'As long as there is a safety valve of unexplored frontiers, then the creative, the aggressive, the exploitive urges of human beings can be channelled into long term possibilities and benefits. But if those frontiers close down and people begin to turn in upon themselves, that jeopardizes the democratic fabric.' The speech was said to have been very like one recently given in San Francisco by the acid poet Timothy Leary, another long-time acquaintance of Brand who was then touring the country promoting the benefits of interplanetary migration.

After the speeches, the event (doubtless thanks to Brand) morphed into a kind of space-age electric kool-acid test as the beat poet Michael McClure read a new work to the accompaniment of some of NASA's most spectacular films. Judging by the photographs and interviews that appeared in *CoEvolution Quarterly*, the social mixture proved very successful. The next day the company moved to Edwards Air Force base for an even more way-out happening: the first (atmospheric) flight test and landing of the new space shuttle. Governor Brown launched a string of proposals: a public communications satellite for the state of California, a new programme of environmental monitoring, and investment in solar power. 'Small is beautiful on Earth, but in space big is better,' proclaimed Brown as the guests drank solar-heated coffee.[45] Brand and his colleagues helped Brown's office produce the *California Water Atlas*, which made use of satellite images to show how this most essential of environmental systems worked.[46]

Russell Schweickart continued to move between space culture and counterculture. After two years at Governor Brown's office, he went on to become chairman of the California State Energy Commission, where, after the Three Mile Island nuclear reactor disaster, he opposed nuclear power, finding common cause with the New Age physicist Fritjof Capra. (Meanwhile his former astronaut colleague Harrison Schmitt as a Republican Senator opposed every form of energy regulation.) At one event Schweickart 'sat in a Pasadena auditorium with a metallic star pasted on his forehead as dancers circled him, chanting for the elimination of nuclear power plants'. From the audience, the prophet Janus of Ora announced: 'Spaceships will be coming this weekend.' Schweickart also got involved with satellite communications companies, the Antarctic programme and efforts to protect the Earth from collision with asteroids, and helped to found the Association of Space Explorers. He compared the experience of seeing the Earth from the outside to leaving the womb and seeing one's mother for the first time: 'I call it the Cosmic Birth Phenomenon.'[47]

Schweickart's peak visibility probably came in 1982 when his 'No frames, no boundaries' meditation was used in a film of the same title, made to promote the Beyond War movement. Beyond War, founded to oppose the nuclear arms race and the 'Star Wars' programme, took the whole Earth as its symbol. The film began with views of the Earth from orbit, with Schweickart's words as the soundtrack. Footage of the edge of the sea was succeeded by film of the Berlin Wall. 'What is a frame? What are boundaries?' asked the commentary. A potted history of mankind traced all this back to the beginnings of agriculture and land ownership, when 'we began building our own frames and boundaries, the genesis of cultural conflict that 10,000 years later would threaten our extinction'.

As images of primitive clubs were succeeded by pictures of nuclear bombs, the commentary warned of the unprecedented devastation that would be inflicted on the planet by nuclear war. Then Schweickart's words were heard again over a picture of the wartorn Middle East from orbit:

There you are. Hundreds of people killing each other over some imaginary line that you're not even aware of, that you can't see. And from where you see it, the thing is a whole, and it's so beautiful. You wish you could take a person in each hand, and say,

'Look. Look at it from this perspective. Look at that. What's important?'

The film was shown, to tremendous effect, at countless public meetings. 'We live on one planet, with one life support system,' it proclaimed. 'The survival of all humanity, [of] all life is totally interdependent.' The last frame showed the Apollo 17 'Blue marble', with the caption, 'The choice is ours.'[48] Only ten years after it had appeared so alive, the Earth seemed confronted by its own mortality.

The second Cold War gave birth to the nuclear winter. Once again, the image of the Earth from space was used to frame the argument. In 1982, the year in which Beyond War was launched, the *New York Times* journalist Jonathan Schell published a small black book called *The Fate of the Earth* about the global consequences of nuclear war. For Schell, the photographs of Earth from space illustrated both 'our mastery over nature . . . and our weakness and frailty in the face of that mastery'. Like an astronaut able to blot out the distant Earth with his thumb, 'as the possessors of nuclear arms we can stand outside nature, holding instruments of cosmic power with which we can blot life out, while at the same time we remain embedded in nature and depend on it for our survival'. In the end, thought Schell, 'the view that counts is the one from Earth, from within life . . . a view not just of one Earth but of innumerable Earths in succession, standing out brightly against the endless darkness of space, of oblivion.' His book, described by Helen Caldicott as 'the new Bible of our age, the White Paper of our time', argued with terrifying realism how nuclear war would destroy first urban civilisation and then the planetary ecosystem, leaving 'a republic of insects and grass'.[49]

Soon after Schell's work was published, Carl Sagan led a group of scientists in an early experiment in planetary climate modelling to work out what would in fact happen if a nuclear war were to pulverise a significant part of urban civilisation and project it into the atmosphere as radioactive debris. The conclusion was that intense global pollution under darkened skies could bring about a prolonged 'nuclear winter'. This would wreck the ecology of the Earth and wipe out not just humanity but all the higher forms of life. Sagan promptly organised a conference on the subject in

Washington DC which produced a much debated book, *The Cold and the Dark*, with contributions from two other whole Earth thinkers, Paul Ehrlich and Lewis Thomas.[50]

The nuclear winter scenario was turned into the darkest of all visions of the whole Earth by the historian and peace campaigner E. P. Thompson. In the early 1980s Thompson interrupted his historical career to attend to the emergency of nuclear disarmament which, he felt, demanded the attention of all who cared about the continuing supply of history. Afterwards he wrote a novel entitled *The Sykaos Papers*, following the ignominious career of a timid and naive alien come to rescue Earth from its follies. Near its end, watchers on the Moon see nuclear war break out on Earth. 'Slowly, so slowly, the blue-and-white globe went brown, shrouding the northern hemisphere, with yellow wreaths extending to the south. Then it was very dim. No blue left, the clouds dun-coloured, some of them black. Around the half-phase rim the refractions seemed red, throwing a red Earthshine on the Moon.'[51] The image of Earthrise had made it possible to envisage the death of the Earth.

It was hardly surprising that in this dark decade the peace and ecology movements failed to share the fascination with the products of rocket technology which had animated parts of the earlier Whole Earth movement. 'Take the toys away from the boys' was the slogan of women peace campaigners. Many of the protestors who displayed the image of the whole Earth had no idea where it came from, or assumed it was just an ordinary satellite photograph; Apollo was forgotten. It was, thought the ecologist Donald Worster, ironic that 'the image of an ailing but ancient organic planet came from the highly polished lens of a mechanical camera carried aloft in a mechanical spaceship'.[52]

The rejection of the technology was accompanied by a more suspicious attitude towards the images which it had generated. When ecofeminist protestors planned direct action at the Nevada nuclear test site on Mother's Day 1987, they wore T-shirts bearing the whole Earth and the slogan 'Love your mother'. The T-shirt image was apt, wrote Donna Hathaway, showing Earth as 'a kind of fetus floating in the amniotic cosmos and a mother to all its own inhabitants ... joining the changeling matter of mortal bodies and the ideal, eternal sphere of the philosophers'. But the use of what she called 'a satellite view of the planet' provided 'a jarring note', recalling 'the space race, and the militarization and commodification of the whole Earth'.[53] The ecofeminist Yakov Gaarb went further: 'As a machine-made representation of a small, external, distant, static object, the whole Earth image is not a suitable symbol for a whole and mutual relationship between humans and the Earth.' As humanity prepared to move into space, 'the Earth is discarded . . . a worn and spent relic from humanity's childhood which can be trashed as we move on.' The whole Earth was now 'a small, comprehensible, manageable icon . . . an impoverished image that symbolises and perpetuates an impoverished world view'. Environmentalists had been conned by technology. This single image, misleadingly undamaged, concealed 'the earthy realities of the biological world'. Gaarb called for 'real and full-bodied images of communion with our planet . . . . Bring on the Sky Goddesses.'[54]

Everywhere now, the whole Earth appeared on advertisements for items as varied as organic beauty products, industrial chemicals and technological hardware. Donna Hathaway felt that the 'bourgeois, family-affirming snapshot of mother Earth' had become 'as uplifting as a loving commercial Mother's Day card', conscripted in support of a social order dominated by 'phallocrats' and 'techno-pornographers'. The feminist writer Ann Oakley voiced the common view that the advertising industry had managed to turn even ecological consciousness into a commodity: 'The appeal to buy "natural" products precisely played on the theme of the Earth as a helpless, beloved woman – a victim to be rescued from degradation by acts of communitarian altruism.'[55] With her naked image displayed across magazine pages and advertising hoardings, Mother Earth seemed to have escaped nuclear destruction only to suffer a fate worse than death.

Thanks largely to the effect of the first views of Earth from space, the Spaceship Earth idea began the 1960s as a technological metaphor and ended it as an ecological one. It was overtaken by the whole Earth movement, which was happy to use technology to raise planetary consciousness, celebrating and cerebrating in equal measure. In the late 1970s and 1980s, whole Earth thinking made most headway in the form of dark fears about the fate of the planet in an age threatened once again by nuclear war; Mother Earth was set in opposition to the technology which had revealed her secrets. At the same time, however, a rather different vision was quietly making headway, in the form of the greatest of all philosophical insights generated by the view of Earth from space: the Gaia hypothesis.

CHAPTER 9

Gaia

'Viewed from the distance of the Moon, the astonishing thing about the Earth, catching the breath, is that it is alive,' wrote the biologist Lewis Thomas in 1974. His words were quoted prominently at the start of James Lovelock's book *The Ages of Gaia*. While working with NASA in the mid-1960s, Lovelock developed the now famous Gaia hypothesis, the most fertile of all the scientific insights that have flowed from the sight of the whole Earth from space. 'The outstanding spin-off from space research is not new technology,' he wrote when introducing the first Gaia book, *Gaia: A New Look at Life on Earth*, in 1979. 'The real bonus has been that for the first time we have had a chance to look at the Earth from space, and the information gained from seeing from the outside our azure planet in all its global beauty has given rise to a whole new set of questions and answers.' Lovelock felt that this also took humankind right back to the ancient concept of Mother Earth: 'Ancient belief and modern knowledge have fused emotionally in the awe with which astronauts with their own eyes and we by indirect vision have seen the Earth revealed in all its shining beauty against the deep darkness of space.'[1]

The Gaia hypothesis has been called 'the most widely discussed scientific metaphor of the Age of Ecology'. As originally formulated, it proposed that 'the Earth's living matter, air, oceans, and land surface form a complex system which can be seen as a single organism and which has the capacity to keep our planet a fit place for life.' More simply, 'the Earth . . . is actively made fit and comfortable by the presence of life itself'.[2] Even more simply, the Earth is alive. Despite initial resistance from sections of the scientific community, Gaia has consistently made headway, going from hypothesis to scientific theory and acting as a model for holistic thinking in the sciences. 'Can there have been any more inspiring vision this century than that of the Earth from space?' exclaimed Lovelock, looking back. 'We saw for the first time what a gem of a planet we live on. The astronauts who saw the whole Earth from Apollo 8 gave us an icon that has become as powerful as the scimitar or the cross.'[3]

The living Earth

'Our recognition of living things, both animal and vegetable, is instant and automatic,' James Lovelock has observed. As early as 1969 the microbiologist René Dubos related how the sight of the Apollo 8 Earth images 'made me realize that the Earth is a living organism'. Eugene Odum, the pioneer of ecosystem ecology, used the Earthrise picture in his textbook *Fundamentals of Ecology* and kept a poster of it on his office wall. Dubos however went further, suggesting that the Earth's natural systems were self-repairing, and the Earth adaptable and resilient. Humankind was part of all this, not separate from it: 'Earth and man are thus two complementary components of a system, which might be called cybernetic, since each shapes the other in a continuous act of creation.' Dubos developed what he called 'a theology of the Earth', arguing that 'a truly ecological view of the world has religious overtones'. Lovelock appreciated Dubos's way of thinking, particularly what he called the 'concept of man as the steward to life on Earth, in symbiosis with it like some grand gardener for all the world'.[4]

While whole Earth thinking in the natural sciences was given a boost by the sight of the Earth, it was not created by it. In the 1950s the palaeontologist and philosopher Loren Eiseley contemplated the ways in which science had come to understand the world as the product of gradual natural processes, rather than of Old Testament style cataclysms. 'Like the body of an animal,' he suggested, 'the world is ... destroyed in one part, but is renewed in another.' Donald Worster has identified a long-standing tradition of 'seeing the Earth as a single living organism', and observes: 'that the Earth was sick, and that the sickness was our doing, was a spreading idea after World War Two'.[5] This tradition had been revived earlier in the twentieth century by the agriculturalist Liberty Hyde Bailey, who wrote: 'The Earth is holy .... We are here, part in the creation.'

Humankind, he felt, had been given dominion but not ownership of the Earth; sadly, 'our dominion has been mostly destructive'. He even wrote of 'the mothership' of the Earth, although he was referring to motherhood rather than spaceships.[6]

In the 1920s the Russian biologist Vladimir Vernadsky had developed the concept of the biosphere. Vernadsky too had been inspired by the view of the Earth from space, although in 1926 he had to imagine it. His book *The Biosphere* opened with these words: 'The face of the Earth viewed from celestial space presents a unique appearance, different from all other heavenly bodies. The surface that separates the planet from the cosmic medium is the *biosphere*.' Defining it as 'the envelope of life where the planet meets the cosmic milieu', he wrote: 'the biosphere plays an extraordinary planetary role . . . the Earth's structure is a harmonious integration of parts that must be studied as an indivisible mechanism.' Lovelock did not know of Vernadsky's work until later; a full English translation was only published in 1998, when he was hailed as 'the first person in history to come to grips with the real implications of the fact that Earth is a self-contained sphere'. The term 'biosphere' went back still further, having been used in the late nineteenth century by the Austrian geologist Eduard Suess. At the start of his book Suess imagined an observer from outer space 'pushing aside the belts of red-brown clouds which obscure our atmosphere, to gaze for a whole day on the surface of the Earth as it rotates beneath him'. He called his book *The Face of the Earth*.[7]

Visual thinkers, it seems, are prone to big ideas. 'When I first saw Gaia in my mind I felt as must an astronaut have done as he stood on the Moon, gazing at our home, the Earth,' wrote James Lovelock, recalling the origins of the Gaia hypothesis.[8] Still at that point no one had seen the whole Earth. The Gaia hypothesis was, however, triggered by the space programme, in what at first seems an unlikely way: by the search for extraterrestrial life.

In the earlier 1960s Lovelock had been employed by NASA to design experiments to detect life on Mars, for a Mars lander to be called Voyager. Lovelock proposed that rather than looking for life directly in the soil, Voyager should look for it indirectly by testing the composition of the atmosphere. His ideas were rejected; so was the Voyager project itself by Congress when a Mars orbiter seemed to show that Mars was all rock and desert. Scientists had to make do with data from Earth-based infrared telescopes, which in 1965 showed that the atmospheres of Mars and Venus consisted mainly of carbon dioxide; there was none of the exotic chemistry found on Earth. These results confirmed Lovelock's hunch that a live planet would have a live atmosphere; as he later put it, 'the dead are more stable than the living.' The health bulletin on Mars, then, was 'stable, but dead'. But if so, what was keeping the Earth's much more complex atmosphere stable and fit for life? Discussing the question with the astronomer Carl Sagan, Lovelock learnt that the Sun had been growing gradually hotter over the lifetime of the Earth, yet he knew that the Earth's temperature range had remained broadly stable. This was Lovelock's eureka moment: 'Suddenly the image of the Earth as a living organism able to regulate its temperature and chemistry at a comfortable steady state emerged in my mind.'[9]

Lovelock's thinking as it emerged from his background in the physical sciences was fleshed out in collaboration with the Boston microbiologist Lynn Margulis, whose research indicated that all forms of life on Earth had developed from the most primitive single-celled organisms. Her grand thesis was that life on Earth had developed not through competition alone but through the relationship between different forms of life, which she called 'symbiotic evolution'. Margulis also had a space connection, having earlier been married to Carl Sagan. Lovelock's suggestion that the composition of the atmosphere was influenced by biological activity fell on deaf ears at a conference at Princeton in May 1968. Back home in his Wiltshire village, Lovelock discussed his concept of the Earth as 'a cybernetic system with homeostatic tendencies' with his neighbour, the novelist William Golding, wondering what to call it. 'I need a good four-letter word,' he mused. Golding suggested 'Gaia', the Greek Earth goddess. The Gaia hypothesis was first published in Carl Sagan's space journal *Icarus* and in the pages of Stewart Brand's *CoEvolution Quarterly*.[10]

Brand's publication was an appropriate home, for the Gaia hypothesis also owed much to the kind of systems thinking promoted by Brand and expressed in Buckminster Fuller's concept of 'Spaceship Earth'. 'Gaia had first been seen from space and the arguments used were from thermodynamics,' wrote Lovelock. 'I found it reasonable to call the Earth alive in the sense that it was a self-organizing and self-regulating system'; his analogy was with an oven controlled by a thermostat. But Lovelock explicitly rejected the systems ecology view that (as Howard Odum had put it) 'the biosphere is really an overgrown space capsule'. He saw this as the 'depressing view of our planet as a demented spaceship, forever travelling, driverless and purposeless, around an inner circle of the Sun'. Only when the population increased to 10 billion or more, he thought, would mankind be reduced to 'permanent enslavement on the prison hulk of the spaceship Earth'.[11]

Space travel and systems thinking also influenced the Gaia hypothesis in another way: by suggesting a model for spreading awareness of the Earth and its problems. 'The evolution of *homo sapiens*, with his technological inventiveness and his increasingly subtle communications network, has vastly increased Gaia's range of perception,' wrote Lovelock towards the end of his book. 'She is now through us awake and aware of herself. She has seen the reflection of her fair face through the eyes of astronauts and the television cameras of orbiting spacecraft.' Perhaps Lovelock had in mind Russell Schweickart's description of himself in orbit as 'the sensing organ for mankind'. In the 1980s Lovelock became a fellow of the Lindisfarne Association, the organisation for 'planetary consciousness' which published and promoted Schweickart's essay. He paid tribute to the 'inspiration and the warm humanity' which he found there, and to the work of its founders, William Irwin Thompson and James Morton. Like Schweickart, Lovelock disliked 'tribalism and nationalism', and he asked: 'Do we as a species constitute a Gaian nervous system and a brain which can consciously anticipate environmental change? Whether we like it or not, we are already beginning to function in this way.'[12]

As well as appealing to the New Age movement, Lovelock also had affinities with the environmental movement. The electron capture device which he designed in 1957 was able to detect traces of man-made chemicals throughout the environment, making possible much of the research that went into Rachel Carson's *Silent Spring* and assisting in the discovery of ozone depletion; appropriately enough, it eventually won Lovelock the Blue Planet prize in 1997. After the initial breakthrough in 1965, his ideas about Gaia were refined during a break from planetary science, investigating the global consequences of air pollution from fossil fuel burning under contract to Shell. Environmentalists, however, did not embrace Gaia as readily as they had embraced the image of the whole Earth. Lovelock's emphasis on the Earth's capacity to repair itself seemed like a feel-good antidote to environmental anxiety, an excuse for complacency, even a licence to pollute, although Lovelock was careful to stress that Gaia's way of recovery would probably not be to humanity's advantage. Lovelock was also critical of the anti-scientific stance of elements of the environmental movement, which in turn found it difficult to accept his further contention that the permanent impact of nuclear power generation, or even of nuclear war, on Gaia would be far less than the cumulative impact of many more ordinary human activities.[13]

Published in book form in 1979, Gaia theory at first encountered what Lovelock called 'a climate of almost religious intolerance' in the orthodox scientific world. The idea of a planet-wide entity capable of controlling itself – an Earth with its own ends, so to speak – was ridiculed by leading biologists, notably Richard Dawkins. 'What an awful name to call a theory,' said John Maynard Smith.[14] Despite its far-reaching consequences, evolution was a process, not an institution; it had no objective, much less any kind of consciousness. In what wider ecosystem had Gaia emerged, and against what competition? From this point of view, Gaia reproduced the errors of other widespread misunderstandings of natural selection, such as social Darwinism, Lamarckianism, neo-creationism, or the mystical ramblings of Teilhard de Chardin. Similar attacks had upset the developing field of systems ecology in the 1960s and 1970s. Lovelock modified his original formulation to avoid the suggestion that Gaia was a purposeful organism, talking of it rather as a set of interlocking systems of which life was one. He persisted, however, with the use of imaginative language which he felt assisted the understanding of concepts. Gaia stayed, and she was still female.

Far from competing with Darwinism, Lovelock believed that Gaia theory added an extra level of sophistication to it. Life did not simply adapt to its environment; seen on a planetary scale, it also adapted the environment to itself. 'Just as the shell is part of the snail, so the rocks, the air and the oceans are part of Gaia,' he wrote, unconsciously echoing Hannah Arendt's anxious forecast about how life seen from space would be as indistinguishable from its surroundings as a snail from its shell. 'The evolution of the species and the evolution of the rocks . . . are tightly coupled as a single, indivisible process.'[15] The atmosphere was a kind of topsoil, recycled and enriched by the organisms that inhabited it; the level of salt in the oceans was kept ideal for life, as if by some kind of global filtering system; the planet even seemed to have its own thermostatic climate control. That was why Earth seen from space looked so different from every other planet. It had not just been sculpted – it had grown.

The Gaia hypothesis, whether by inspiring assent or disagreement, encouraged holistic thinking about the Earth in a number of other areas, from the study of climate change to the search for extraterrestrial life. A widespread acceptance developed that some living systems could 'eco-engineer' their own environments. Gaia found itself riding – and indeed leading – the trend towards the integration of the sciences which had been gathering strength ever since the more or less simultaneous advent of ecology and space science. New disciplinary labels began appearing: Lovelock's 'geophysiology', NASA's 'Earth systems science', and even 'astrobiology'. In 2001, over a thousand delegates from four international organisations met in Amsterdam to consider how to construct 'an ethical framework for global stewardship' and 'a new system of global environmental science'. The declaration they signed was regarded by Lovelock as a vindication: 'The Earth system behaves as a single, self-regulating system comprised of physical, chemical, biological and human components.'[16]

Gaia was generally welcomed in the New Age and environmental movements for its emphasis on human responsibility for the Earth, on the capacity of modern technology to upset the ancient balance of Mother Earth, and on the need for a general raising of awareness. It informed the eco-feminism of the 1980s and 1990s. Lynn Margulis had decidedly mixed feelings about the confused idea of Gaia in popular culture. 'The planet takes care of us, not we it. Our self-inflated moral imperative to guide a wayward Earth or heal our sick planet is evidence of our immense capacity for self-delusion. Rather, we need to protect ourselves from ourselves.' Far from being a female entity in need of rescue, 'Gaia, a tough bitch, is not at all threatened by humans. . . . Our tenacious illusion of special dispensation belies our true status as upright mammalian weeds.'[17]

The Gaia hypothesis also challenged the inhibition in some parts of the religious world against believing that the created could ever really harm the Creation and so defeat the purposes of God. As Mother Teresa put it, 'Why should we take care of the Earth when our duty is to the poor and sick among us? God will take care of the Earth.'[18] There was also the usual suspicion of all monotheistic religions towards anything that smacked of pantheism and nature worship. The kind of evangelical neo-conservatism that welcomed the apocalypse, and which helped to legitimate the climate change denial of the Bush presidencies, had no room for Gaia. But Lovelock's ideas sat well within another important Christian tradition, that of the human duty of stewardship towards the creation.

Soon after the first Gaia book came out, Lovelock was contacted by James Morton, co-founder of the Lindisfarne Association and also dean of the cathedral church of St John the Divine in New York – the famous spire among the skyscrapers. He persuaded Lovelock, following the example of other Earth thinkers such as William Irwin Thompson and René Dubos, to give a sermon: 'it was a sensual experience,' Lovelock recalled. Dressed in robes, he ascended the pulpit as the congregation sang 'Morning has broken' and then spoke about the newly published Gaia hypothesis. The Bishop of Birmingham, Hugh Montefiore, was enthralled by the book and wrote to him, asking provocatively: 'Which came first – God or Gaia?' Montefiore's book *The Probability of God* spread awareness of Gaia among Christians. Lovelock's second book, *The Ages of Gaia*, included a whole chapter on 'God and Gaia', giving encouragement to this new Gaian-style strand of Christianity.[19]

The Gaian account of the ways in which life interacted with its environment, and of the ways in which humans were disrupting long-established planetary mechanisms, did much to accustom public opinion to the idea that humankind could indeed do serious damage to the Earth. The ground for the acceptance of the reality of global warming was prepared by the Gaia hypothesis. From the Gaian perspective, global warming is common sense. Over hundreds of millions of years, Earth's living systems have kept the composition of the atmosphere and the temperature of the Earth low and stable by absorbing and then burying carbon and methane, much of it in the form of coal, gas and oil – what are called 'fossil fuels'. If in the course of a few generations a substantial part of Gaia's safely interred waste products are dug up, burnt and dumped back into the atmosphere, there is bound to be trouble. If at the same time half the vegetation that the Earth needs to filter the poison from its system is stripped off and itself burnt, then drastic action is needed. The indicator is the extent of polar ice; for the North Pole, at least in summer, that looks likely to be gone within the lifetime of many of those reading this book, and with it the chance of ever regaining the Earth we once knew.

All this is visible from space. Space shuttle astronauts in the 1980s who had also been into orbit in the 1960s thought that the colours of the Earth no longer seemed quite as rich. 'The brilliant, clear photos were the Gemini photos of the mid-1960s,' observed Richard Underwood. 'The air pollution was a lot less then, and it shows.'[20] The 1972 'Blue marble' photo, which showed the relatively undeveloped southern hemisphere, was notable for its view of the Antarctic in winter; already things look different. There have been no journeys like it and no photos like it since. Our first full-face portrait of the Gaia that lived then may also turn out to be the last.

Climate change

By the late 1980s, most people in the western world were at least vaguely aware that something was going wrong with the planet. Thanks to the International Geophysical Year of 1957–8, which had launched programmes of both polar and space research, the combined efforts of satellite monitoring, atmospheric research and Antarctic surveys had revealed a large and growing hole in the ozone layer over the Antarctic. This threatened the integrity of the Earth's shield against ultraviolet radiation. An international meeting at Vienna in 1985 recognised the problem, and another followed at Montreal in 1987. The minds of the delegates were concentrated by dramatic pictures of the hole taken by NASA's Nimbus satellite in the autumn of 1986. The result was a series of agreements regulating and eventually banning CFCs, the aerosol and refrigeration gases which were causing the problem. As with Earth Day in 1970, a view of the Earth from space had prompted environmental action, this time at a global level. Within a few years the measures appeared to be working, setting a hopeful precedent for dealing with the much more intractable problem of global warming.

As if in sympathy with the wounded Earth, in 1987–8 there appeared three compelling books placing the view of Earth from space firmly before the public once again. Frank White's *The Overview Effect*, with its compelling astronaut interviews, was followed by a superbly produced anthology of American astronaut photography, *The View from Space*. At a time of serious doubt over the value of the human presence in space, this book for the first time really brought out the photographic skills of the astronauts themselves.[21] Russell Schweickart followed up his involvement in the Beyond War movement by helping to found the Association of Space Explorers (ASE) which (as we saw in Chapter 6) had its first conference in 1985. By 1988 he was reported to be a divorced father of five, living on a houseboat in Sausalito, California, not far from the founder of the *Whole Earth Catalog*, Stewart Brand. In October 1988 Schweickart toured the United States with the cosmonaut Alexander Alexandrov to promote ASE's book about the Earth, *Home Planet*, perhaps the most evocative of all books about the Earth.

In October 1989, as the population of Eastern Europe began breaking through the Iron Curtain and crowds gathered at the Berlin Wall itself, the ASE held a second conference about the Earth, this time in Saudi Arabia. 'We . . . have seen from afar the dangers that can imperil our planet,' the delegates declared. They called for an international network of environmental monitoring stations, the use of global communications to 'bring nations and peoples closer', and more international cooperation in space.[22] That autumn, as the Iron Curtain unexpectedly dissolved, humanity began to mingle once more along the political boundaries which had seemed so meaningless from space.

As the Cold War gave way to global warming, 1990 saw a twentieth anniversary repeat of the first Earth Day. This second major Earth Day was more mainstream than the first, with much more involvement from private companies and much less in the way of militant action. This was not surprising in view of the progress that environmentalism had made; a poll showed that 76 per cent of Americans now considered themselves environmentalists – with even more among those who were white and better off, though fewer among the very rich. NASA chose the occasion to announce its Mission to Planet Earth programme, a fifteen-year $20 billion-plus programme for an Earth Observing System, to include at last two satellites in polar orbits. Pictures from the orbiting space shuttle continued to keep the state of the planet in the public eye; there could no longer be any doubt that rainforests were burning, deserts advancing and polar ice caps retreating. A new Landsat followed. After a failed launch in 1993, from 1999 Landsat 7 was able to photograph the Earth's entire surface every sixteen days.[23]

In 1992, two years after the revived Earth Day, the United Nations staged an even bigger twentieth anniversary environmental event: the monumental Rio Earth Summit. This was the successor to the UN conference on the Human Environment at Stockholm in 1972, which had been inspired by the first whole Earth photographs. Ever since then the UN's environmental commission had been steadily working to keep the planetary perspective represented within the realm of practical politics. Its 1987 report *Our Common Future* declared:

> From space, we see a small and fragile ball dominated not by human activity and edifice but by a pattern of clouds, oceans, greenery, and soils. Humanity's inability to fit its doings into that pattern is changing planetary systems, fundamentally. From space we can see and study the Earth as an organism whose health depends on the health of all its parts.[24]

Following this report, the Intergovernmental Panel on Climate Change was established in 1988, and in 1992 delegates from 172 countries and (crucially) numerous non-governmental organisations met in Rio de Janeiro at the Earth Summit. Some of them travelled in a Viking-style ship named *Gaia*, which went from Norway via Port Canaveral in Florida. A deputation gathered there to wave it off on the second stage of its journey, within sight of the massive launch towers which had once sent the Apollo astronauts on their way. The speakers were James Lovelock and James Lovell, scientist and astronaut, who compared notes about the Earth. Lovelock explained how the Gaia hypothesis had its roots in the space programme. Lovell explained how he had come to understand that 'home was the planet', and showed how from the Moon his thumbnail, held at arm's length, could cover up the Earth completely. As they talked, remembered Lovelock, 'the small vessel set sail for Rio, taking with it its message of a living Earth'.[25]

The idea of climate change was strongly resisted in some quarters. It posed the same threat to the prospect of unlimited progress on Earth that the view of Earth posed to the prospect of unlimited progress in space: it showed that there were limits. After several days of debates, planetary in scope, the Earth Summit issued the Rio Declaration on environment and development. It proclaimed that 'states shall cooperate in a spirit of global partnership to conserve, protect and restore the health and integrity of the Earth's ecosystem'. It also set out an action programme towards the next century, known as Agenda 21, covering a huge range of global environmental issues from greenhouse gases to endangered species. This in turn led five years later to the Kyoto Protocol on climate change, an agreement which in its first decade had little measurable effect on the problem but which was of groundbreaking political significance.

The challenge to national sovereignty posed by environmental movements was an important element in what Jay Winter regards as the twentieth century's last 'utopian moment' in 1992 (or, to be less exact, in the years 1989–92): a wave of activity around the issue of global citizenship.[26] The delegates to the Rio Earth Summit recognised the need to supplement (or perhaps short-circuit) the activities of national governments by involving citizens everywhere in local environmental actions. Over the next few years Agenda 21 items became a routine feature of meetings of local authorities and citizens' forums the world over. This practical experiment in transnational citizenship stemmed directly from the transnational nature of the environmental problems which it sought to address. It was an appropriate development for an environmental movement which had first come together on an Earth Day inspired by the sight from space of a world without boundaries and symbolised by a flag bearing the image of the whole Earth.

Pale blue dot

While all this was going on, new and more distant photographs of the Earth were being taken. The means to take them had been in existence for some time, in the form of a deep space probe called Voyager which in 1977 had set out to explore the solar system, equipped with a camera to observe the planets as far out as Neptune – an object so faint and distant that it had not been detected until 1846. In the year that Voyager set out the cosmologist Steven Weinberg published a popular book on the origins of the universe, *The First Three Minutes*. His closing statement caused a stir. 'It is almost irresistible for humans to believe that we have some special relation to the universe,' he wrote, 'that human life is not just a more-or-less farcical outcome of a chain of accidents reaching back to the first three minutes, but that we were somehow built in from the beginning.' But however inviting the Earth might appear from on high, this was all an illusion. 'It is very hard to realize that this all is just a tiny part of an overwhelmingly hostile universe . . . [which] faces a future extinction of endless cold or intolerable heat. The more the universe seems comprehensible, the more it also seems pointless.'[27]

This bleak vision worried other cosmologists, including Carl Sagan, a member of the Voyager team. Sagan's life was devoted to conveying to the widest possible public the excitement and magic of the universe, offering what has been called 'a mythic understanding of science'. In his immensely popular book of the 1980 television series *Cosmos* he offered an implicit reply to Weinberg. 'We are the local embodiment of a cosmos grown to self-awareness. We have begun to contemplate our origins. . . . Our loyalties are to the species and the planet. We speak for Earth. Our obligation to survive is owed not just to ourselves, but also to that Cosmos, ancient and vast, from which we sprang.' 'The inescapable recognition of the unity and fragility of the Earth', he wrote later, 'is . . . the unexpected final gift of Apollo.' But he also felt that a still more distant photo 'might help in the continuing process of revealing to ourselves our true circumstance and condition. It had been well understood by the scientists and philosophers of classical antiquity that the Earth was a mere point in a vast encompassing cosmos, but no-one had ever *seen* it as such.'[28]

Sagan's own gift to humankind would be a picture of the Earth taken from the outer reaches of the solar system. Voyager had in fact taken a picture of the Earth and Moon together early in its journey. He first made his proposal when Voyager passed Saturn in 1981, but because of the risk of burning out the camera by pointing it towards the Sun it was put off until the probe had passed Uranus. Even then there were arguments within the Voyager project; some thought that such a picture just 'wasn't science'. With budgets being cut and key technicians about to be laid off, the head of NASA had to step in to settle the dispute before it was too late; he came down in Sagan's favour. It was now February 1990 and the probe was passing the orbit of the outermost planet, Neptune. After careful calculations of angles and exposure times, Voyager photographed six of the eight planets. 'This is how the planets would look to an alien space-ship approaching the solar system after a long interstellar voyage,' explained Sagan. Earth was a pretty blue, but so were Uranus and Neptune. From four billion miles out Earth seemed tiny and inse-cure, but nothing special. It was, in Sagan's phrase, a 'pale blue dot'.

'Look again at that dot,' he urged, echoing the words of Russell Schweickart. 'That's home. On it everyone you know, everyone you ever heard of, every human being that ever was, lived out their lives . . . on a mote of dust suspended in a sunbeam.' Voyager also took the Earth in another sense. On board the space probe as it headed out of the solar system was a package of images and sounds chosen to represent the essence of life on Earth: music, voices, landscapes, the cry of a baby, the voice of a whale, portraits of people of all races, a photo-graph of a Gemini astronaut spacewalking above the blue ocean, and another of the Earth from space. One day, imagined Sagan and his colleagues, somewhere else in the galaxy, intelligent aliens might inter-cept Voyager, open it up, and gaze upon a picture of the whole Earth.[29]

In 1992, the year of the Earth Summit, another remarkable picture arrived from deep space, more striking because more recog-nisable. The Galileo project had been conceived after the Voyager launch in 1977 and it was launched from the orbiting space shuttle in 1989. Picking up speed on its way to Jupiter, it twice swept by the Earth, and on the second pass, in December 1992, from 4 million miles out, it photographed the Earth and the Moon together. Even more than other pictures of the whole Earth, it conveyed a three-dimensional sense of a planet floating in space, similar to what the astronauts had experienced. It was a pebble rather than a disc. It was also a family photo rather than a portrait, except that one member was obviously stone dead and the other brilliantly alive.

As the 'Pale blue dot' image arrived, Carl Sagan led a coalition of prominent humanist scientists in writing 'an open letter to the religious community'. Freeman Dyson, Stephen Jay Gould and Lynn Margulis were among the thirty-two signatories. 'We are close to committing – many would argue we are already committing – what in religious language is sometimes called "crimes against creation",' warned the letter. The problem 'must be recognised as having a religious as well as a scientific dimension. . . . Efforts to safeguard and cherish the environment need to be infused with a vision of the sacred.' Religious people of all faiths were asked 'to commit, in word and deed, and as boldly as is required, to preserve the environment of the Earth'. Numerous religious leaders hastened to rise to the challenge and the result was the Joint Appeal by Religion and Science for the Environment, headed by Sagan himself and James Parks Morton, New York's 'green dean' (who had privately prompted Sagan's original open letter). 'The cause of environmental integrity and justice must occupy a position of utmost priority for people of faith,' it declared. The National Religious Partnership for the Environment was established, bringing together Protestants, Catholics, Jews and others in a move to build in an ecological dimension to faith traditions – the 'greening of religious faith'.

A regular Gaian religious ceremony was begun at Morton's New York cathedral. On the festival of St Francis of Assisi, people were encouraged to bring their pets to a special service. The 'green dean' made a real effort to involve scientists in the ceremony, regardless of religious faith, in order to emphasise the spiritual dimension to the study of nature. 'Since God is dead, we do biology,' said one. Morton would recite a christening liturgy written by a biologist, talking of how 'life is born of water, and born again of water in each day's turning of the Earth'. Then the congregation hushed as a procession of living things was solemnly led through the quietened church: a plant, a boa constrictor, an eagle, a camel, even an elephant, and always a glass sphere filled with blue-green algae, representing the Earth. Connie Barlow, an environmental scientist, was among those present. 'The Dean . . . talks about the trillions of beings alive in that glass bowl and how they are among the first organisms that lived on this planet. . . . I hear through the hush kids whispering, "Here come the algae! Here come the algae!" '

Connie Barlow wrote an Earth Day invocation for her own church. 'In our hearts we call forth gratitude for that air that swirls around and within us, for the ancient microbes that two billion years ago discovered how to crack a molecule of water, feeding on the hydrogen and setting the oxygen free – forever changing the orange-brown sky of primordial Earth into a brilliant blue.' She was part of an informal network of environmental scientists who wanted to contemplate the wider meaning of Gaia. 'When I go for a walk in the woods now or just look up at the clouds, I feel like I'm inside a giant metabolism,' said one. Barlow also wrote her own credo: 'The evolution epic is my creation story, and this self-renewing living planet is creation's greatest achievement.'[30]

Among those who took part in the green dean's dialogues between science and religion was the Democratic Senator Al Gore. Gore was closely involved with the UN's recently formed Intergovernmental Panel on Climate Change. To promote its work he wrote a book, *Earth in the Balance*, which appeared in 1992. Illustrated by the Apollo 17 view of Earth, the book called for a 'global Marshall plan' to save the environment. A few months later Al Gore became US Vice-President. Among the initiatives he announced in office was to make an 'all Earth, all-the-time' video available as part of NASA's Mission to Earth programme. As he did so he quoted Russell Schweickart's 'on that small spot' speech, carrying on the hybrid environmental/space tradition developed by Schweickart, Brand and Governor Jerry Brown of California in the 1970s.[31]

In 2000 Gore ran for president, losing to George W. Bush in the closest presidential race in American history. Given what we now know about the urgency of tackling global warming, and the timely difference which US support might have made to forging an effective global consensus, more than was apparent at the time hinged on the outcome of ballot irregularities in Florida, the state that had launched Apollo.

America forfeited a green president but the world gained a green advocate. Gore toured extensively with a compelling slide show on climate change, lecturing in front of a huge image of the Apollo 'Blue marble'. Picture after picture of the Earth, from ground level and space alike, enabled his audiences to see as if they were astronauts, viewing the evidence of environmental damage for themselves: pollution haze, burning rainforests and oil wells, soil erosion, ozone depletion and collapsing ice shelves. In 2006 the slide show became a film and a book, *An Inconvenient Truth*. It opened with the Earthrise photograph and an account of the voyage of Apollo 8, quoting from the Genesis broadcast and Archibald MacLeish's 'Riders on the Earth'; next came the 'Blue marble'. It ended with the same two pictures, and then Carl Sagan's 'Pale blue dot'. Polls showed American citizens to be even further ahead of their own government in appreciation of global warming than those of other nations. A year later, Gore and the Intergovernmental Panel on Climate Change shared the Nobel Peace Prize.

The year 2007 also saw the UN climate change conference in Bali, where 12,000 delegates assembled in an almost millennial atmosphere of anxiety and expectation in order to plan the next global agreement on climate change after Kyoto. For days the US government delegation did all it could to block and frustrate consensus, and by the final exhausting all-night session even experienced UN negotiators were at breaking point. The climax was marked by a shaming moment in American history as the delegate from Papua New Guinea rounded on the world's only superpower, telling it: 'If you're not willing to lead, get out of the way.' A few minutes later the US delegation gracefully gave way, bowled over from behind by a tidal wave of environmental feeling which had circled the world after being set off in America nearly forty years before by the release of the first Earthrise picture.[32]

Rare Earth

At the end of the twentieth century, further indications of the uniqueness of the Earth came from an unexpected quarter: astrobiology, the study of life on other planets – a subject in search of an object if ever there was one. The presumed existence of intelligent life elsewhere in the universe had been a fundamental tenet of astrofuturism, underpinning its core assumption about the relative insignificance of life on Earth. The belief that man was the only intelligent form of life was regarded as a relic of the pre-Copernican period when human beings regarded Earth as the centre of the cosmos. Acceptance of the possibility of extraterrestrial intelligence was, on the other hand, regarded as the natural culmination of 'the slow realization that our planet does not occupy a central position'. 'Nowadays, the writing for the old life-originated-here-on-the-Earth prejudice is on the wall', wrote the astronomer Fred Hoyle in 1980.[33]

The first technologically equipped searches for intelligent life beyond the solar system got underway almost as soon as the first space age began, with Frank Drake's 'Project Ozma' at the US National Radio Astronomy Observatory in 1959–60. The quest produced the famous 'Drake equation', which suggested that there were likely to be 10 million communicating civilisations in our galaxy alone (give or take the odd zero). It was not exactly an equation but rather a shopping list in mathematical notation of all the necessary assumptions: 'a way of compressing a large amount of ignorance into [a] small space', according to one critic. The Drake equation was an imaginative way of overcoming the basic impossibility of extrapolating from a sample of one. Quickly popularised, however, Drake's views were by 1966 regarded as an orthodoxy.[34]

The film *2001: A Space Odyssey*, made in 1964–8 in response to Drake's ideas, dramatised the common assumption that humanity's first steps into space would quickly be followed by contact with other civilisations, benign and God-like. 'I believe that what they have to tell us is of supreme importance. I feel certain we can find them now,' wrote Drake in 1991. 'It is mathematically certain . . . I believe I shall still be alive when we'll know,' said the ageing space pioneer Willie Ley. 'This planet will die, will grow cold just as it grew warm, and we must make ready to leave.' International lawyers, in a very large volume published by Yale University Press, urgently pressed for a system of 'law and public order in space' in order to create 'a commonwealth of dignity for all advanced forms of life'.[35]

Project Ozma drew a blank, and so did a raft of other attempts to scan the skies, collectively known as the Search for Extra-Terrestrial Intelligence (SETI); Carl Sagan was a leading advocate. By the mid-1970s even supporters began asking out loud Enrico Fermi's pointed question: 'If extra-terrestrials exist, where are they?' Theorists were at a loss to explain 'the great silence'; some lost their faith, including Sagan's Russian co-author Iosef Shklovskii (who had once put the number of habitable planets in the galaxy as high as 1 billion). SETI continued, but as its search power multiplied the results counter remained stuck at zero. Against expectations, Earth was beginning to look special once again.[36]

Meanwhile, Earth scientists from a variety of disciplines were finding that Earth had a much more eventful history than had previously been supposed. The 1960s saw the rapid rise to acceptance of the theory of plate tectonics – continental drift. And while the Moon rocks brought back by Apollo did not immediately settle the debate about the origins of the Earth and Moon, by the mid-1980s strong evidence had emerged that they were formed when a gigantic object hit the primeval Earth at an early stage of the formation of the solar system. Discoveries about the Earth made from on high through the space programme, combined with a holistic way of thinking that owed much to the Gaia hypothesis, produced an outline history of life on Earth that challenged space age assumptions about the ordinariness of the blue planet. This work was rounded up in a book called *Rare Earth* by Peter Ward and Donald Brownlee, a geologist and an astronomer, which in January 2000 (according to one reviewer) 'hit the world of astrobiologists like a killer asteroid'.[37]

Life, suggested Ward and Brownlee, was quite likely widespread in the universe at the level of microbes. It had probably originated on Earth some 4 billion years ago, quite soon after the heavy bombardment by asteroids stopped. But complex life was another matter entirely, and the authors listed all the unusual circumstances of Earth that made it possible: 'not in the centre of a galaxy, not in a metal-poor galaxy, not in a globular cluster, not near an active gamma ray source, not in a multiple-star system, not even in a binary, or near a pulsar, or near stars too small, or too large, or soon to go supernova'. It would also have taken at least two complete star life cycles somewhere else in the galaxy to create the rare heavy elements which life seemed to require. On Earth, at least, while bacterial life had brewed up pretty quickly, complex animal life had required another 3 to 4 billion years of relative stability, orbiting round a warm and well-behaved star, on a planet with just the right amount of water and other building materials, just enough heat to get plate tectonics working, and a single good-sized Moon to act as a gravitational shepherd. Alter any one of these factors and humans would probably not be here to tell the story. 'If animals are as rare in the Universe as we suspect,' explained the authors, 'it puts species extinction in a whole new light. Are we eliminating species not only from our planet but also from a quadrant of the galaxy as a whole?'[38]

Humankind now appears to be both the product and the custodian of the only island of intelligent life in the knowable universe. The astronauts' revelation that the Earth was the only point of life and colour in the infinite blackness of space now seems more significant than ever. Whether that vision has been timely enough, and powerful enough, for homo sapiens, the most successful of all invasive species, to reverse its own devouring impact on the Earth, will probably become apparent before too long.

CHAPTER 10

The discovery of the Earth

Neil Armstrong remarked that from the Moon, the Earth was so small that he could blot it out with his thumb. Did this make him feel big, he was asked. 'No,' he replied, 'it made me feel really, really small.'[1] The questioner assumed that Armstrong would identify with the view from the Moon; Armstrong, however, identified with the Earth. There, in a nutshell, is the story of Earthrise.

Interestingly, Armstrong also became the lunar traveller who most completely rejected the role of the hero and sought to return to Earth as a private citizen. He was not alone, but there were few like him because there were so few to start with. Only twenty-four people have ever seen the whole Earth for themselves, all of them between December 1968 and December 1972. Most of them survive; all retired from space travel long ago. Unlike those celebrities who measure their achievements by the small cultural world which they inhabit, those who have seen the whole world do not appear to be given to vanity or arrogance. In the 2007 film *In the Shadow of the Moon* many of these Moon voyagers appear, not acting as tribunes of progress or transformed into apostles of some new cosmic awareness, but as ordinary Earthmen, looking back with maturity and insight on the extraordinary events of their youth. They seem more than usually grateful to be living on the Earth.

This was not quite what had been predicted earlier. Celebrating Apollo 11, the well-known astronomer Patrick Moore wrote: 'We have entered a new age. Earthly isolationism ended at that moment on July 21, 1969, when Neil Armstrong and Buzz Aldrin stepped out on to the Sea of Tranquillity.' The astronauts, however, felt equally the pull of space and that of the Earth as seen from space. The man who took the 'Blue marble' photograph, Harrison Schmitt, wrote, 'like our childhood home, we really see the Earth only as we prepare to leave it . . . we seek more than the beauties and comforts displayed on the rising Earth. Paradoxically, we become more aware and appreciative of that which is so willingly left behind.' Addressing Congress after returning from the Moon, Michael Collins described how the space capsule had slowly rotated:

> As we turned, the Earth and the Moon alternately appeared in our windows. We had our choice. We could look toward the Moon, toward Mars, toward our future in space, toward the New Indies, or we could look back toward the Earth our home, with the problems spawned over more than a millennium of human occupancy. We looked both ways. We saw both, and I think that is what our nation must do.[2]

Astrofuturists, poised in their imagination between the Earth and the universe, had difficulty looking both ways at once. Frank White, whose book *The Overview Effect* explored the impact of the sight of the Earth, sought to reconcile the inward and outward visions. He coined a word for those who achieved astronaut-type awareness without going into space, 'terranauts', and argued that once about 20 per cent of the population achieved such a change of consciousness it would rapidly spread to the rest. For White, the overview effect was the foundation for 'a series of new civilizations evolving on Earth and in space'. Similarly Wyn Wachorst, in the astrofuturist elegy *The Dream of Spaceflight*, worried that the view of Earth had driven humanity inwards. He compared 'the sight of our fragile, lonely world' to 'the new self-awareness that comes when one realizes that the parent is mortal'. It was a stage in maturity, part of the process of leaving home and becoming independent.[3]

Carl Sagan, as both an astrofuturist and an environmentalist, felt the Earth/space dilemma acutely and perhaps came closer than anyone to resolving it, but at the cost of relegating the exploration of the stars to the far future. In his last popular book, *Pale Blue Dot*, he considered the significance of the Voyager photo: 'Our posturings, our imagined self-importance, the delusion that we have some privileged position, are challenged by this point of pale light. Our planet is a lonely speck in the great enveloping cosmic dark. . . . There is no hint that help will come from elsewhere to save us from ourselves.' In the far future, thought Sagan, humans would travel to the stars and overcome 'the sting of the Great Demotions', but until then, 'this distant image of our tiny world . . . underscores our responsibility to deal more kindly with one another, and to preserve and cherish the pale blue dot, the only home we've got'.[4]

One name was often invoked in these discussions: that of Copernicus. The first Earth photographs were widely expected to hammer home the Copernican awareness that the Earth was not the centre of creation but just a planet among others. The problem was that in the pictures, Earth remained at the centre of the frame, appealing so eloquently direct to the viewer that the astronomical context was almost forgotten. In the early twenty-first century, whatever people's intellectual grasp of cosmology, the general level of environmental awareness suggests that the Earth is as much at the centre of human attention as ever.

A similar story emerges if we look at literature. In the aftermath of the Apollo programme the Aerospace Industries Association sponsored a study of the cultural impact of space exploration, and was disappointed to find that it had had 'minimal influence on literature, especially poetry'; the poets' response was simply 'years of silence'. Surveying the literature, Ronald Weber found that 'rather than the view into space, it is the view from space leading back to Earth and its ordinary concerns that has occupied literary minds'. 'With this image, the literary imagination turned homeward and inward,' concluded Valerie Neal of the National Air and Space Museum. The persistent theme of the few poets who wrote about the space programme was the contrast between the dead Moon and the live Earth, and the human emptiness of space travel. For Robert Phillips, the Moon offered 'Lots of rocks . . . . Lots of dust . . . . But nothing like home.' Archibald MacLeish's lunar explorers in 'Voyage to the Moon' stood blankly on her beaches, finding meaning only by looking back at the Earth. The Russian poet Andrei Voznesensky wrote a poem about Mars and called it 'Earth':[5]

> Somewhere on Mars
>   A visitor from Earth
> Will take out a handful of warm, brown Earth

And lovingly gaze at the blue-green sphere,
Never distant,
Ever near!

Those who sought literary quotations to enhance books about the space programme often ended up quoting lines from T. S. Eliot's 'Little Gidding' about how the end point of all exploring was to know your home for the first time. Scott L. Montgomery, surveying the whole run of western literature about the Moon, found a single thread running through it all: 'Copernicus was, in some sense, wrong after all: the Earth has never ceased being the centre of the universe.' Wachorst wistfully summed up the effect of Apollo: 'The mythic Moon . . . was brought down to Earth while the Earth was placed in the heavens and given the name Gaia.'[6]

It is pleasant to play with ideas, but how can the view of the whole Earth be set in historical context? The three main stages in seeing the whole Earth – curving horizon, Earthrise/'Blue marble' and 'Pale blue dot' – coincide neatly with the three 'utopian moments' which Jay Winter has identified since the Second World War: 1948, the year of the Universal Declaration of Human Rights; 1968, the year of youth rebellion; and 1992, the year of global citizenship. All of them were also associated with flurries of space-related idealism. In 1948–50 there was the Berlin airlift, the first well-known pictures of part of the Earth globe and the first popular manifestos for space travel. In 1968–72 came the Apollo programme, the 'Blue marble', the first Earth Day and the first Earth Summit. The years 1989–92 saw the end of the Cold War, the 'Pale blue dot', the revival of Earth Day, and the second Earth Summit. In between these three utopian moments lay two periods of Cold War, moderated by a decade of detente that began with Apollo. Just as the Second World War was followed by the freedom of the air, so the most dangerous phase of the nuclear arms race was followed by the freedom of space. In its internationalist aspects, the space programme was part of the unfinished business of the Second World War.

Until the space age, people had to guess what the Earth would look like. There was a debate about whether the land would appear bright and the seas dark, or vice versa. As late as 1950, Fred Hoyle went for dark sea, though (like others before him) he expected that the sunlight would dazzle as it reflected off the oceans.[7] While a rare visionary like Vernadsky anticipated that the Earth would appear uniquely alive, it seems to have been generally expected that if the Earth was a planet like any other, it would look like any other. The diagram published to show the flight plan of Apollo 8 showed the Earth and the Moon differing in size, but broadly similar in style and colour, twins in style if not in size. Until the 1960s, artists depicting planet Earth all imagined it pretty much the same way: the land was green and brown and dominated the view, the sea was an even blue, and fluffy clouds adorned but did not obscure the familiar outlines of the continents. In short, they showed a geographical globe floating in space on a sunny day.

When the first widely publicised V-2 photographs of the curving Earth showed (as the publicity had it) that 'Copernicus was right', they didn't change this basic geographical view. It was certainly exciting to see the Earth curving away, even as a dome, but the viewer's attention was directed to the details: here was the Gulf of California, there was the Rio Grande, and so on. It was like being unable to see the wood for the trees. This effect was heightened by the use of black-and-white film with an infrared filter to penetrate the haze, and by the rocketeers' habit – so irritating to meteorologists – of always launching in good weather over New Mexico. People could see that the Earth was a globe, but they knew that already, and over the 1950s photos that simply showed more and more of it aroused little response. The early satellite pictures were not that different, and nor were the early black-and-white pictures from the Mercury programme. Then Gemini 4 brought back the first stunning pictures of the blue planet in the spring of 1965, some of them taken from outside the spacecraft. This was at last the real Earth, and as it was compared to traditional globes and maps the absence of political boundaries aroused comment.

From orbit it was one world; from a distance it was the whole Earth. The high-orbit satellite pictures of the Earth that followed, however, aroused less interest than the Gemini pictures, perhaps because they were less attractive; they had a flat, disc-like appearance and a slightly artificial electronic colouring, and they were presented as planet-sized weather maps. The first Earthrise, the black-and-white one from Lunar Orbiter, was different again, for it showed the Earth as seen from the Moon. But the film was black and white and

the resolution poor. The picture seemed like an astronomical arte-
fact, and NASA sought to present it like that, with the lunar horizon
vertical and the Earth at one side, but it ended up printed horizontally all the same. It excited space enthusiasts precisely because it was
not like a view of home.

The Apollo 8 Earthrise was something else again. The rough
black-and-white television pictures of the Earth on the way out had
moved minds, but the sight of a coloured pebble hovering above the
barren lunar landscape moved hearts. The framing was important.
Pictures of the home planet alone were blown up by NASA to fill
the frame, but framed as a landscape shot there was a lot of space
around the rising Earth. Earth was seen in deep time as well as in
deep space: the view seemed eternal, like a snapshot of the Creation.
For astrofuturists, the space around it was the point; for everyone
else, it was the Earth.

Earthrise could be taken in two ways: either as a revolutionary
new perspective of the Earth as a base for space exploration, or as an
evocative portrait of home. Either way, one thing was obvious to all:
as the biologist Lewis Thomas put it, while the Moon was 'dead as
an old bone', the Earth was 'the only exuberant thing in the cosmos'.[8]
One of the most popular songs of 1969, 'Wandrin' star', was an elegy
for the life of the old-time frontier pioneers, but it could equally
have been about the pioneers of the space frontier. 'I've never
seen a sight that didn't look better looking back', ran one line. Its
double-edged message of wanderlust and homesickness captured
Earthrise perfectly.

The Earthrise photograph of 1968 and the 'Blue marble' photograph of 1972 between them frame the Apollo Moon programme.
They also represent the beginning and the summit of whole Earth
awareness. But while Earthrise showed the Earth in space, 'Blue
marble' showed the Earth alone. Filling the frame, centred on Africa
(mankind's place of origin), and looking both alone and alive, its
message was not 'space' but 'home'. It was a record of a particular
historical moment: mankind's last trip (to date) beyond Earth orbit.
'Blue marble' was mankind's holiday postcard to itself; like many
holiday postcards, it arrived back after those who sent it.

For Carl Sagan, wearing his cosmologist's rather than his environmentalist's hat, the 'Blue marble' photograph 'conveyed to multitudes

something well known to astronomers: on the scale of worlds – to say nothing of stars and galaxies – humans are inconsequential, a thin film of life on an obscure and solitary lump of rock and metal'.[9] For most Earth dwellers, however, the view of humankind's home from a distance was significant not because of the distance but because it was home. Rather than hovering behind the shoulder, Earth was the target of the camera, humbling but not humiliating, a view of the future and not of the past. The photo that became the most widely reproduced image in human history was also like the most common type of picture in the world: a full face portrait. The subject was Gaia. To the father who would leave his home and family to voyage in space, it gazed back reproachfully as if to say: 'Look, this is your child.'

The effect of any photograph of Earth is limited by its context as a photograph: around it, the world remains the same. But, as Michael Collins explained, it was quite different 'to actually be 100,000 miles out, to look out four windows and find nothing but black infinity, to finally locate the blue-and-white golf ball in the fifth window . . . one must get 100,000 miles away from it to appreciate fully one's good fortune in living on it'.[10] Astronauts got a sense of a pebble-like planet adrift in three-dimensional space which is experienced on Earth only by those fortunate enough to witness a total eclipse of the Moon on a clear night. Perhaps, thought space advocates, the Copernican message had not sunk in as expected because the Apollo pictures did not carry the full Copernican hit. Carl Sagan's 'Pale blue dot' of 1990, and the Galileo Earth–Moon photo of two years later, conveyed this three-dimensional sense much better, partly because of the sheer amount of blank space in the frame. The Galileo photograph, with the Earth and Moon floating together, the one in front of the other, is most effective. The 'Pale blue dot', however, remains just that: a dot. Like the Earth as imagined by ancient philosophers, it is an aid to thought; it does not impress as it does from close at hand.

The fascination that Earthrise continued to exert was shown by the careful photography plans made for the Japanese Space Agency's lunar probe Kaguya, which journeyed gently to the Moon in the autumn of 2007. On board was a pair of high definition TV cameras – one wide angle, one telephoto, pointing in opposite directions. The Apollo craft had all adopted approximately equatorial orbits (horizontal in Earth terms), which had meant that the Earth rose at one side of the Moon, in practice usually at a slight angle. The timing of these missions to get the right kind of light for Moonwalks meant that the full Earth was never seen from the Moon. Kaguya by contrast has adopted a polar (vertical) orbit, so that the Earth rises and sets straight over the lunar poles. Twice a year, when Sun, Moon and Earth all line up, the Earth rises as a completely full disc – although as the probe is coming from under the Moon's south pole at the time, the Earth appears upside down. Kaguya arrived a little too late to capture a full Earthrise in the autumn of 2007, but sent back two brilliantly detailed pictures of a nearly full Earthset (telephoto) and Earthrise (wide angle). The deep lunar shadows created a quite different feeling from the Apollo photographs. In April 2008, as this book was going to press, Kaguya flipped right round to bring the telephoto lens to the Earthrise position and on 6 April captured Earthrise, crisp and full.

Kaguya's Earthrise was the perfect image at the perfect moment – a haiku in picture form. It was a technological achievement but also a cultural one. At the same time it feels (at least to some western sensibilities) almost too digitally perfect. It illustrates just how much the Apollo 8 Earthrise gained from the knowledge that there was a human eye behind the camera.[11]

As the Earth was first shown to its inhabitants, its parish boundaries, so to speak, were extended. Until the first space age they lay well within the atmosphere; travel into orbit and beyond was equated with the first step out into the cosmos. Many expected that entry into space would be followed before too long by contact with extraterrestrial intelligence, and with it a revolution of the conditions of human existence; the search for extraterrestrial intelligence represented a kind of high-tech cargo cult. Soon, however, orbit came to seem like an extension of airspace. Within a generation, satellites, shuttles and space stations became a routine part of life on Earth, beaming down continuous information, helping to keep the world's population in touch, in place, and entertained.

While Earth's boundaries have expanded to include orbital space, the cosmos has grown vast beyond all imagining. During the Apollo years and after, the discovery of pulsars, quasars and cosmic background radiation pushed the physical boundaries of the universe virtually to infinity and its age back to some 13 billion years. In 1995 the Hubble space telescope took a long hard look at a tiny patch of space and sent back a deep field photograph of one quarter of a degree of sky. It showed three thousand galaxies, their light warped by gravitational lensing, seen as they were when the universe was young. The gap between the Earth and rest of the universe seems as vast as ever, in time as well as space. No one waves back. In 1992 NASA's COBE telescope produced a map of the cosmic microwave background radiation, the echo of the big bang. It is a view from inside the sphere of the cosmos, turned inside out and projected onto two dimensions. We now have not only a picture of the whole Earth but a picture of the whole universe.

Interviewed nearly forty years after the voyage of Apollo 8, James Lovell was pessimistic about its enduring effects. 'We didn't know the impact of the flight on the people of the world until we got back. . . . But the mind forgets very easily, and not too long after that people got back to the way they lived before – wars and disruption and human cruelty. People don't realize what they have until they leave, and only a few people have done that.' This view is hard to argue with; perhaps it isn't really one world after all. But if we consider the impact of the 'whole Earth' perspective, perhaps rather more has changed.[12]

The sight of the whole Earth, small, alive, and alone, caused scientific and philosophical thought to shift away from the assumption that the Earth was a fixed environment, unalterably given to humankind, and towards a model of the Earth as an evolving environment, conditioned by life and alterable by human activity. This view was most clearly expressed by Gaia theory, itself a product of the first space age. Its career corresponded to Schopenhauer's dictum that 'all truth passes through three stages: first, it is ridiculed; second, it is violently opposed; and third, it is accepted as self-evident'.[13] Acceptance of the idea that the conditions of life on Earth were changed by the activities of living systems had become general by the early twenty-first century, in parallel with acceptance of the fact of climate change. The sight of the whole Earth gave the world a picture to think with.

All this amounted to a paradigm shift, along the lines of Thomas Kuhn's model of scientific revolutions.[14] It was a relatively rapid one. It would be interesting to compare the speed and scope of this shift with that which accompanied the discoveries of the Renaissance age of exploration (which, unlike the first space age, did find new varieties of intelligent life). The discovery of the New World was succeeded nearly five centuries later by the rediscovery of the Earth.

Among future-minded thinkers, the sight of the distant Earth was anticipated as a sign that humankind was growing up and leaving home at last. It had also been anticipated that the first steps in space would be followed before long by contact with extra-terrestrial intelligence, confirming the demotion of Earth. Yet within a few years doubts began to be heard; the great silence that greeted the first phase of SETI also suggested that life (or at least civilisation) as found on Earth was rare indeed. The Earth was assumed to be receding; then came Earthrise. Those views of the whole Earth, the space programme's greatest trophy, were also its greatest liability, for they conveyed to the public the implicit message that humanity's much-vaunted 'destiny in space' was in fact the rediscovery of the home planet. Astrofuturists had forgotten one crucial thing: everyone on Earth could face outwards, but only astronauts could face the Earth.

The further people have travelled from the home planet, and the more the universe around it has expanded, the more precious and unique the Earth has appeared. Out of all the contradictions of the space programme came a vision beyond price: from the Moon, mankind first saw the Earth. The flight of Apollo 8 was the climax of the modern age; still, no-one has travelled further or faster. Within four years the manned lunar programme was over. Forty years later its greatest legacy is that miraculous picture to which history has supplied the caption: 'Earthrise, seen for the first time by human eyes'. The space programme, which was meant to show mankind that its home was only its cradle, ended up showing that its cradle was its only home. It was the defining moment of the twentieth century.

Notes

References to NASA (National Aeronautics and Space Administration) for Apollo materials are to files consulted at the Johnson Space Center (JSC), Houston (formerly called the Manned Spaceflight Center); the files have since been transferred to Rice University. Much of this material is also at the NASA History Office (NHO), Washington DC. Where files may be unique to one site or the other, I have specified either 'JSC' or 'NHO'. References to JSC OHA are to the JSC Oral History Archive available at http://www.jsc.nasa.gov/history/

Chapter 1 Earthrise, seen for the first time by human eyes

1. United Nations Treaty on Principles Governing the Activities of States in the Exploration and Use of Outer Space, including the Moon and Other Celestial Bodies (1967), quotation from Article 5.
2. Apollo 8 onboard voice recorder transcript, JSC, at 75h 47m.
3. Frank Borman, *Countdown: An Autobiography* (New York, 1988), 212.
4. Frank Borman interview, JSC OHA, 1999; *Life*, 17 Jan. 1969.
5. Bill Anders interview, PBS Nova TV documentary *To the Moon*, 1999, transcript in JSC. The text of the programme (though not the whole interviews) is available at the website of the US Public Broadcasting Service, http://www.pbs.org/wgbh/nova/tothemoon/; Tony Reichardt, 'Leaving Home', *Final Frontier* (December 1988).
6. William Irwin Thompson, 'The deeper meaning of Apollo 17', *New York Times*, 1 Jan. 1973. Cousins repeated his thought in a symposium at CalTech in July 1976, called to mark both the American bicentennial and the success of NASA's Viking Mars Lander: *Why Man Explores*, NASA EP 125, 1976.
7. Asif Siddiqi has pointed out that in the original Russian, this phrase translates as 'the Earth is the cradle of reason, but one cannot live in a cradle forever'. But the version almost universally quoted in the West is the one that is relevant for our purpose. Asif A. Siddiqi, 'American space history', in Roger Launius and Steven Dick (eds), *Critical Issues in the History of Spaceflight*, NASA SP 2006–4702 (Washington DC, 2006), 435n.

8. De Witt Douglas Kilgore, *Astrofuturism* (Philadelphia, 2003).
9. William E. Burrows, *This New Ocean: The Story of the First Space Age* (New York, 1998), 143 and ch. 2; Dennis Piszkiewicz, *Wernher von Braun* (Westport, CT, 1998); Howard McCurdy, *Space and the American Imagination* (Washington, DC, 2002), chs 2–3; Roger Launius, 'Perceptions of Apollo: myth, nostalgia, memory, or all of the above?', *Space Policy* 21 (2005); Bruce Mazlish (ed.), *The Railroad and the Space Programme: An Exploration in Historical Analogy* (Cambridge, MA, 1965).
10. Roger Launius and Howard McCurdy, *Imagining Space* (San Francisco, 2001), 2, 5.
11. McCurdy, *Space and the American Imagination*, 29, 58 and chs 2–3; R. Liebermann, 'The *Collier's* and Disney series', in Frederick W. Ordway and Randy Liebermann (eds), *Blueprint for Space* (Washington, DC 1992); Launius and McCurdy, *Imagining Space*.
12. Launius, 'Perceptions of Apollo'; Mazlish, *The Railroad and the Space Programme*. Mazlish's project, funded by NASA 'in an enlightened and farsighted mood' (said its editor), was introduced by the space specialist of the American Association for the Advancement of Science, who found it 'illuminates a number of parallels between the railroad and the space effort, thus providing for NASA and the American people a perceptive guide to the effects of peaceful space exploration and their meaning for American society'.
13. Arthur C. Clarke, 'The challenge of the spaceship', *Journal of the British Interplanetary Society* (Dec. 1946), developed in 'Space flight and the spirit of man', *Astronautics* (Oct. 1961), and reprinted in Clarke's *Voices from the Sky* (London, 1966), 3–11; Ronald Reagan, address to National Space Club luncheon, 29 Mar. 1985, quoted in John Logsdon, 'Outer space and international policy: the rapidly changing issues', in D. S. Papp and J. R. McIntyre (eds), *International Space Policy* (Westport, CT, 1987), 31.
14. Robert Poole, '*2001: A Space Odyssey* and the space age', *History Today* (Jan. 2001).
15. Paine Papers, Library of Congress (LOC), box 43.
16. Associated Press bulletin, 24 Dec. 1968; *Miami News*, 24 Dec. 1968; *Evening Star* (Washington), 26 Dec. 1968.
17. *Look*, 4 Feb. 1969.
18. Apollo 8 post-recovery press conference, 27 Dec. 1968, JSC 078/11.
19. *Evening Star* (Washington), 23 Dec. 1968; *Los Angeles Times*, 29 Dec. 1968; *Time*, 3 Jan. 1969; *New York Times*, 28 Dec. 1968; *Congressional Record*, 9 Jan. 1969.
20. Paine to 'Shap' [Willis Shapley], 25 Oct. 1968, Paine Papers, LOC, box 52, folder 7; *Houston Chronicle*, 28 Dec. 1968.
21. *Washington Post*, 28 Dec. 1968; *Chicago Daily News*, 28 Dec. 1968; *New York Times*, 28 Dec. 1968; *Evening Bulletin* (Philadelphia), 26 Dec. 1968; *St. Louis Globe-Democrat*, 28–29 Dec. 1968.
22. *Christian Science Monitor*, 27 Dec. 1968, and see chapter 7 below.
23. *Life*, 10 Jan. 1969. The poem is Dickey's 'For the first manned moon orbit', part of a suite of space poetry: James Dickey, *The Central Motion: Poems 1968–1979* (Middletown, CT, 1983), 22–6.

24. 'Riders on the Earth' *New York Times*, 25 Dec. 1968.

25. *Sunday Denver Post*, 29 Dec. 1968; *Christian Science Monitor*, 27 Dec., 28–30 Dec. 1968.

26. Daniel C. Noel, 'Re-entry: Earth images in post-Apollo culture', *Michigan Quarterly Review* 17 (1979), 173–4; René Dubos, 'A theology of the earth', lecture at Smithsonian Institution, 2 Oct. 1969, published in revised form in *A God Within* (1972; London, 1976); John Noble Wilford, 'The spinoff from space', *New York Times* magazine, 29 Jan. 1978.

27. John Caffrey, quoted in Oran W. Nicks (ed.), *This Island Earth*, NASA SP-250 (Washington, DC, 1970), 3–4; Donald Worster, *Nature's Economy* (1977; 2nd edn Cambridge, 1994), 358–9; Galen Rowell in *Sierra*, Sept. 1995, 73, quoted in Zimmerman, *Genesis: The Story of Apollo 8* (New York, 1998), 284.

28. Andrew Smith, *Moondust: In Search of the Men Who Fell to Earth* (London, 2005), 283–7. Smith's book goes some way in this direction, as does Frank White, *The Overview Effect* (1987; 2nd edn, Reston, VA, 1998). Both are based on astronaut interviews. White's book explores the profound impact which the sight of the Earth from space had on the astronauts themselves, and argues that 'the overview effect and its related phenomena are the foundations for a series of new civilizations evolving on Earth and in space'. Smith's book recounts more recent interviews with those who had walked on the Moon. Neither book looks specifically at the perspective of the whole Earth. White treats orbital and distant views of the Earth together, while Smith interviews the nine surviving astronauts who had walked on the Moon but not the twelve who went there but didn't land.

29. Michael Light, *Full Moon* (1999; London, 2002).

30. McCurdy, *Space and the American Imagination*; Burrows, *This New Ocean*; Roger Launius, 'The historical dimension of space exploration: reflections and possibilities', *Space Policy* 16 (2000); Launius, 'Perceptions of Apollo'; Launius and Dick (eds), *Critical Issues*, particularly the overview essays by Stephen J. Pyne, Asif Siddiqi and Roger Launius, wherein may be found references to 'the new aerospace history'. Burrows seems to have been the earliest to write of 'the first space age', although he extends it almost to the present. Two fine pieces of writing examine space travel in relation to nature and humanity: Anne Morrow Lindbergh, 'The heron and the astronaut', *Life*, 28 Feb. 1969, and Kenneth Brower, *The Starship and the Canoe* (1978; London, 1980).

31. There is also the problem that a number of astronaut memoirs are ghosted, or co-written with a professional writer. I am not too bothered about this. A comparison of the Apollo 11 astronauts' rushed memoir, *The First Men in the Moon*, produced by a writer on *Life* magazine, with the individual memoirs produced by Michael Collins (*Carrying the Fire*) and by Buzz Aldrin with Wayne Warga (*Return to Earth*) suggests that real ghosting is pretty obvious and that the assistance of professional writers does not significantly distort memoirs that go out under an astronaut's own name.

32. Jay Winter, in *Dreams of Peace and Freedom: Utopian Moments in the Twentieth Century* (New Haven, CT 2006), picks 1968 as one of his six 'utopian moments', but strange to say Apollo 8 does not figure in it.

Chapter 2 Apollo 8: from the Moon to the Earth

1. Michael Collins, *Carrying the Fire: An Astronaut's Journey* (New York, 1975), 475; Michael Collins interview, JSC OHA, 1997.

2. Asif A. Siddiqi, *Challenge to Apollo: The Soviet Union and the Space Race, 1945–1974*, NASA SP-2000 4408 (Washington, DC, 2000); Deborah Cadbury, *Space Race* (London, 2005); Gene Krantz, *Failure is Not an Option* (New York, 2000), 226.

3. Reginald Turnill, *The Moonlandings: An Eye Witness Account* (Cambridge, 2003), 151.

4. Samuel Phillips, 'Apollo 8 decision' transcript, 12 Nov. 1968 (NASA). Unless otherwise stated, information and quotations about the course of the Apollo 8 mission are from NASA's transcripts of the mission at JSC 078-11, particularly the flight plan of 22 November 1968, the press kit of 15 December, the Apollo 8 mission commentary, the onboard voice recorder transcript and the summary log. Many of these are now available at the NASA History Office website, http://history.nasa.gov. Original film, pictures and commentary are available on the DVD *Apollo 8: Leaving the Cradle* (Fox, 2002).

5. Samuel Phillips, 'The shakedown cruises', in Edgar Cortright (ed.), *Apollo Expeditions to the Moon*, NASA SP-350 (Washington, DC, 1975), ch. 9; Deke Slayton (with Michael Cassutt), *Deke! US Manned Space: From Mercury to the Shuttle* (New York, 1994), 212–17. C.G. Brooks, J.M. Grimwood and L.S. Swenson, *Chariots for Apollo: A History of Manned Lunar Spacecraft*, NASA SP-4205, 1979.

6. Frank Borman, *Countdown: An Autobiography* (New York, 1988), 188. He gave the same account in an interview for the 1999 PBS Nova documentary, *To the Moon,* which presents further evidence, and added in a JSC Oral History Archive interview in 1999: 'we really weren't certain that the Russians weren't breathing down our backs, so I wanted to go on time.' Lovell has also written that 'the CIA told us the Rusians were aiming to circumnavigate the Moon', *Observer*, 20 Jan. 2008, review section p. 3.

7. Slayton, *Deke!*, 212–17.

8. Ibid., 216–17; Siddiqi, *Challenge to Apollo*, 662; Phillips, 'The shakedown cruises'. Although some involved were convinced that intelligence about Soviet intentions helped to drive the decision, hard evidence is elusive. W. H. Lambright, in *Powering Apollo: James E. Webb of NASA* (Baltimore, 1995), 196–200, writes of a 'strong feeling, based on intelligence information, that the revitalized Soviet program might at this very moment be preparing for such a flight'. The inconsistency in the story is that Webb, who would have had the highest access to intelligence, was the least enthusiastic about the decision; he resigned on 15 September for other reasons, removing perhaps the main block.

9. Paine memo to James R. Jones, 30 Oct. 1968, Paine Papers, LOC, box 52, folder 7; Paine to Helms, 4 Dec. 1968, and to Mueller, Maugle and Beggs, 31 Jan. 1969, NHO file 011085.

10. See chapter 5 below. As late as July 1969 Apollo 11 was shadowed to the Moon by a Soviet craft, Luna 15, believed to be capable of carrying crew, which passed as close as 10 miles (16 km) to the Moon and (it later turned

out) crashed while attempting to get a surface sample. The excitement caused to television viewers by the suggestion that Soviet cosmonauts were secretly aloft is still remembered. David Scott and Alexei Leonov, *Two Sides of the Moon* (London, 2004), 225–7; Turnill, *The Moonlandings,* 132–4; Siddiqi, *Challenge to Apollo,* 653–68, 694–6.

11. Krantz, *Failure is Not an Option,* 226–7. This first use of the overworked phrase '101 per cent' was in fact correct: Apollo 7 had fulfilled every target, and one extra added at the last minute. The phrase caught on in the footballing world, and devaluation rapidly set in.

12. Bill Anders interview, JSC, *To the Moon,* PBS Nova TV documentary, 1999.

13. Krantz, *Failure is Not an Option,* 238–40; Collins, *Carrying the Fire,* 304.

14. Lindbergh, 'The heron and the astronaut'; *Baltimore Sun,* 10 Nov. 1967, commenting on an earlier Saturn V launch; Susan Borman, interviewed for 'The astronauts' wives' story', BBC Radio 4, Nov. 2007; Norman Mailer, *A Fire on the Moon* (London, 1970), 82. Mailer was commenting on the similar Apollo 11 launch.

15. Michael Collins interview, JSC OHA; Collins, *Carrying the Fire,* 304–5; Krantz, *Failure is Not an Option,* 242.

16. Piers Bizony, *The Man Who Ran the Moon* (Cambridge, 2007), 215–16; *Life,* 17 Jan. 1969, 27; Borman, *Countdown,* 204.

17. Krantz, *Failure is Not an Option,* 215, 228. The Apollo 8 onboard computer had a memory of some 40,000 words, half the length of this book, of which only 2,000 words were erasable.

18. Apollo 8 onboard voice transcription, JSC 078-13; *Life,* 17 Jan. 1969, 29.

19. Personal communication, June 2007.

20. *The Listener* (London), 20 Feb. 1969; Borman, *Countdown,* 209–11.

21. French and Burgess, *In the Shadow of the Moon,* 310–12.

22. Zimmerman, *Genesis*; the 1999 PBS Nova TV documentary *To the Moon*; Richard Underwood to Glenn Swanson, JSC History Office, 23 Dec. 1998; JSC 077-65 Apollo 8 Photographic TV and Operations Plan, 18 Nov. 1968.

23. Billy Watkins, *Apollo Moon Missions: The Unsung Heroes* (Westport, CT, 2006), an account put together from a series of telephone interviews, 30–1.

24. Frank Borman, personal communication, June 2007; French and Burgess, *In the Shadow of the Moon,* 310–12.

25. The preparations for this broadcast are described in chapter 7 below.

26. Krantz, *Failure is Not an Option,* 245.

27. William Styron, Foreword to Ron Schick and Julia van Haften, *The View from Space: American Astronaut Photography 1962–1972* (New York, 1988).

28. Collins, *Carrying the Fire,* 59–61.

29. *New York Times,* 28 Dec. 1968; *Honolulu Advertiser,* 28 Dec. 1968. The pilot of the inbound flight, which had come from Australia via Fiji, had been given the coordinates before leaving and changed course to give the passengers a better view. 'All of us pilots see quite a bit of space garbage . . . but I have never seen anything like this,' said the pilot. The first sight of the returning Apollo 8, however, was picked up as early as 26 December, in England, through a telescope by pupils at Austin Friars Grammar School, Carlisle, Cumbria. 'We picked up Apollo very clearly as the Moon darkened slightly at 7:25 p.m. There

is no doubt at all about it being Apollo. We have been waiting since the launch to spot it,' said Father Dedan Deasy, their teacher (*The Times*, 27 Dec. 1968).

30. Collins, *Carrying the Fire*, 312; *New York Times*, 28 Dec. 1968; Krantz, *Failure is Not an Option*, 246.

31. Richard Underwood interview, 17 Oct. 2000, NASA Oral History Archive, see pp. 22–4, at http://history.nasa.gov. Underwood's role in preparing the astronauts to photograph the Earth, including those of Apollo 8, is discussed in chapters 4–5 below.

32. Watkins, *Apollo Moon Missions*, 30–1.

33. The wording is copied from an original Public Affairs Office released photo preserved by Bill Larsen at the JSC astronaut office.

34. Copy in NHO.

35. John Catchpole, 'A question of viewpoint', *Spaceflight* 40 (June 1998), 221–3, and subsequent correspondence in *Spaceflight* (Aug. 1998), 286; *Analysis of Apollo 8 Photography and Visual Observations*, NASA SP-201 (1969), 124–5, 188–9; Underwood interview, JSC, 6–7; Zimmerman, *Genesis*.

36. Robert Zimmerman, 'Photo finish', *The Sciences* (Nov.–Dec. 1998), 16–18. See also Catchpole, 'A question of viewpoint'; H. J. P. Arnold, 'Apollo 8 earthrise images', *Spaceflight* (June 1999); *Analysis of Apollo 8 Photography*.

37. Paine to James Jones, Special Assistant to President, 26 Dec. 1968, Paine Papers, LOC, box 22; Paine to Charles H. Townes, University of California, 20 Jan. 1969, Paine Papers, LOC, box 23; Morrow to Paine, Jan. 1969, Paine Papers, LOC, box 43.

38. *New York Times*, 25 Dec. 1968; *Evening Star* (Washington), 30 Dec. 1968; *Sunday Star* (Washington), 29 Dec. 1968. The *Sunday Star* was making a comparison with the atomic bombing of Hiroshima, which was not directly seen at the time.

39. *Time*, 3 Jan. 1969; *Life*, 17 Jan. 1969; *National Geographic*, May 1969; *Washington Post*, 30 Dec. 1968; *Houston Chronicle*, 27 Jan. 1969; *The Times*, 31 Dec. 1968, 6 Jan. 1969. The only serious note of dissent came from the International Flat Earth Society, whose president, Samuel Shenton, had been approached by enterprising pressreporters. On Christmas Eve he had appeared open-minded about the prospect of photos of the Earth: 'If they show us a very clear picture of the Earth from space, and the picture does not show all the continents, and the edge of the world is out of perspective, then that would prove that the Earth is round.' But a few days later he was denouncing them as 'blatantly doctored. Studio shots, probably,' or lens distortion, or a conspiracy of doomed globe-makers. 'We've lost a lot of members because of this absurd Apollo trip,' he complained. Shenton's reaction to the Lunar Orbiter photograph had been almost identical: he felt it at first as 'a blow to the belly', then as 'fraud, fake, trickery, or deceit', *New York Times*, 25 Dec. 1968; *Washington Post*, 4 Jan. 1969; Christine Garwood, *Flat Earth: The History of an Infamous Idea* (London, 2007), 246–8 and ch. 7.

40. *Los Angeles Times*, 29 Dec. 1968.

41. NASA Daily News Bulletins, Jan. 1969; *The Times*, 21 Jan. 1969; Borman, *Countdown*, 223–5; NHO files on Borman and on Apollo 8.

42. Frank Borman, in *This Week*, 6 Apr. 1969; *The Times*, 2 Feb. 1969 onwards; *Washington Post*, 12 Feb. 1969; *The Listener* (London), 20 Feb. 1969, 232–3. On the parallels between Apollo 8 and Verne's novel, see chapter 3 below.

Two years later, Borman was presented in Paris with copies of the Tintin books *Destination Moon* and *Explorers on the Moon* by Hergé: *The Sun* (Baltimore), 11 Aug. 1971.

43. *Sunday Bulletin* (Philadelphia), 16 Feb. 1969; Borman, *Countdown*, 231. On papal interest in space travel, see chapter 7 below,

44. Frank Borman, in *This Week*, 6 Apr. 1969; Frank Borman biographical file 000212, NHO.

45. *The Times*, 27 Dec. 1968; *Washington Post*, 9 Feb. 1969; Paine Papers, LOC, box 43.

46. *Evening Star* (Washington), 2–6 July 1969; *Current Digest of the Soviet Press*, 30 July 1969. Borman did not visit the Baikonur launch pad where an N-1 moon rocket had blown up in a colossal explosion on the eve of his visit: Siddiqi, *Challenge to Apollo*, 688–93.

47. *Life*, 17 Jan. 1969; Russell Schweickart, 'No frames, no boundaries', in Michael Katz, William P. Marsh and Gail Gordon Thompson (eds), *Earth's Answer* (New York, 1977), 2–13.

Chapter 3 A short history of the whole Earth

1. *Universe* (National Film Board of Canada, 1960). There is a copy in the Smithsonian Institution's National Air and Space Museum (NASM).

2. Arthur C. Clarke, *2001: A Space Odyssey* (New York, 1968), ch. 47; Zimmerman, *Genesis*, 57–8; *New York Times*, 28 Dec. 1968. 'An interesting movie, but did not have much effect on me,' recalled Lovell nearly forty years later (personal communication).

3. Denis Cosgrove, *Apollo's Eye: A Cartographic Genealogy of the Earth in the Western Imagination* (Baltimore, 2001), ix, xii, 3. This section draws upon Cosgrove's important and fascinating work.

4. Fred Hoyle, *The Nature of the Universe* (Oxford, 1950), iii, 9–10; entry for Peter Laslett, *Oxford Dictionary of National Biography*; Jane Gregory, *Fred Hoyle's Universe* (Oxford, 2005), 46–9; BBC Written Archives Centre RCONT 1 (Fred Hoyle talks 1947–62), LR/50/217, 427 and 1636 (1950). The original script is identical to the published version, except that the words 'through a blue filter' are added before the description of Earth. Laslett's marginal notes indicate particular interest in the point about space travel adding a physical dimension to human awareness, and scepticism about the suggestion that the sight of Earth will expose the futility of nationalistic strife.

5. Russell Schweickart, 'Earth: Planet 3A of Sol', *Bell Rendezvous* (spring 1970).

6. Arthur C. Clarke, 'If I forget thee, O Earth', in Clarke, *Expedition to Earth* (New York, 1953); Robert Heinlein, *The Green Hills of Earth* (Chicago, 1951), originally published 1947 in the *Saturday Evening Post*; Scott and Leonov, *Two Sides of the Moon*, 314.

7. Arthur C. Clarke, 'Maelstrom II' (1962), in Clarke, *The Wind from the Sun* (London, 1974), 18–19.

8. Arthur C. Clarke, *The Exploration of Space* (1951; 2nd edn, London, 1958), 182; Clarke to C. S. Lewis, Dec. 1943. Bodleian Library, Oxford, Eng. Lett. c220/4 item 8.

9. 'Space flight and the spirit of man', *Astronautics* (American Rocket Society) (Oct. 1961), reprinted in Clarke, *Voices from the Sky*, 3–11. In a footnote to this piece, Clarke later distanced himself from Toynbee.

10. Archibald MacLeish, 'The image of victory', in Hans W. Weigert and Vilhjalmur Stefansson (eds), *Compass of the World: A Symposium on Political Geography* (London, 1946), 1–11. MacLeish's essay was first published in *Atlantic Monthly*, July 1942.

11. Weigert and Stefansson, *Compass of the World*; Robert Wohl, *The Spectacle of Flight: Aviation and the Western Imagination 1920–1950* (New Haven, CT, 2005).

12. Richard E. Harrison and Hans W. Weigert, 'World view and strategy', in Weigert and Stefansson, *Compass of the World*, 74–88; MacLeish, 'The image of victory', 5–7.

13. Winter, *Dreams of Peace and Freedom*, Introduction.

14. Quoted in William Sims Bainbridge, *The Spaceflight Revolution* (New York, 1976), 150; the passage is quoted in full in chapter 8 below.

15. US Department of State press release, 9 July 1965. Stevenson was addressing the UN Economic and Social Council in Geneva on the subject of international development.

16. Yuri Melvill, 'Man in the space age', *Soviet Life* (May 1966).

17. Kilgore, *Astrofuturism*, ch. 1, quotations at 41.

18. Arthur C. Clarke, 'Memoirs of an Armchair Astronaut (Retired)', *Journal of the British Interplanetary Society* 46 (1993), 411–14; Robert Crossley, *Olaf Stapledon: Speaking for the Future* (Liverpool, 1994), 364–5.

19. Olaf Stapledon, *Star Maker* (1937; London, 1972), 16–18, 257–62; Olaf Stapledon, *The Opening of the Eyes* (London, 1954), e.g. 43–5 ('On the threshold of the Church I am spending my life in doubt. . . . The heavens declare – nothing'). On Vernadsky, see chapter 9 below.

20. Konstantin Tsiolkovsky, *Beyond the Planet Earth*, trans. Kenneth Syers (1916; London, 1960), ch. 9 and passim.

21. Cosgrove, *Apollo's Eye*, 205.

22. Winter, *Dreams of Peace and Freedom*, 12, 26–7, 206, and ch. 1 passim. Three of Winter's six 'utopian moments' – 1900, 1948 and the early achievements of the United Nations, and 1968 – were associated with visions of the whole Earth, although only in the first does Winter expound the link. Selections from Kahn's visual archive of the planet were shown in the parochially titled BBC series *Edwardians in Colour* (2007).

23. H. G. Wells, *The First Men in the Moon* (London, 1901), ch. 5.

24. Jules Verne, *Autour de la Lune* (1870; Paris, 1978), 38–40; Borman, *Countdown*, 225; James Lovell interview, JSC OHA, 1999.

25. Edgar Allan Poe, 'Hans Phaall: a tale' (1835), in F. K. Pizor and T. A. Comp (eds), *The Man in the Moone and Other Lunar Fantasies* (New York, 1971), 174–5, and in Poe, *The Complete Tales and Poems of Edgar Allen Poe* (New York, 1938), 30–1; 'The landscape garden', in Poe, *Complete Tales and Poems*, 609; Peter Nicholls (ed.), *The Encyclopedia of Science Fiction* (London, 1979), entry under 'Poe'.

26. Cosgrove, *Apollo's Eye*, 205.

27. C. F. Volney, *The Ruins, or Meditations on the Revolutions of Empires* (New York, 1890), 14–15.

28. Steven J. Dick, *The Biological Universe: The Twentieth Century Extra-terrestrial Life Debate and the Limits of Science* (Cambridge, 1996), 515–16.

29. Here I paraphrase from Cosgrove, *Apollo's Eye*, chs 6–7.

30. David Cressy, 'Early modern space travel and the English man in the moon', *American Historical Review* 111(5) (Dec. 2006); Francis Godwin, 'The man in the moone', in Pizor and Comp, *The Man in the Moone*, 21–2; Scott L. Montgomery, *The Moon and the Western Imagination* (Tucson, 1999), 145–6. Godwin died in 1633 so his account was written some years before its publication in 1638 and can have owed nothing to Kepler.

31. John Wilkins, *The Discovery of a World in the Moon* (1638; Amsterdam, 1972), 148–50 and propositions 8 and 11 generally. This curious idea suggests that Wilkins had not yet properly absorbed Copernican astronomy.

32. Johannes Kepler, *The Dream* [*Somnium*] (1634; Berkeley, 1965), 104–6, 139–41; Dick, *Biological Universe*, 515.

33. 'Monsieur Auzout's speculations of the changes, likely to be discovered in the Earth and the Moon, by their respective inhabitants', *Philosophical Transactions* (Royal Society) I (1665), 120–2; and see below, chapter 9.

34. Cosgrove, *Apollo's Eye*, 120.

35. Peter Whitfield, *The Image of the World: Twenty Centuries of World Maps* (London, 1994), 52–3; R. W. Shirley, *The Mapping of the World: Early Printed World Maps 1472–1700* (1984; Riverside, CT, 2001), 45.

36. Cosgrove, *Apollo's Eye*, 106–10. It is not known whether Ptolemy made a globe and none of his maps survive.

37. Francisco D'Ollanda, *De aetatibus mundi imagines*, facsimile edn (Lisbon, 1983), plate 5 and p. 279.

38. Jeffrey Burton Russell, *Inventing the Flat Earth* (New York, 1991); Edward Grant, *Planets, Stars and Orbs: The Medieval Cosmos 1200–1687* (Cambridge, 1994), 619–22, 626; Whitfield, *The Image of the World*, vii.

39. Cosgrove, *Apollo's Eye*, 50–2, 267, quoting Cicero, *De republica* (London, 1928), 269.

40. *Lucian* trans. A. M. Harmon (London, 1960), vol. 2, 287–8, 299–300; Cosgrove, *Apollo's Eye*, 3.

41. Cosgrove, *Apollo's Eye*, 3–5, 29–34.

42. Carl Sagan, 'A pale, blue dot', *Washington Post/Parade Magazine*, 9 Sept. 1990, 13–15; Harrison Schmitt, 'The new ocean of space', *Sky and Telescope* (Oct. 1982), 327–9.

Chapter 4 From landscape to planet

1. Beaumont Newhall, *Airborne Camera* (London, 1969); N. A. Rynin, *Interplanetary Flight and Communication*, vol. 2, no. 4: *Rockets* (Leningrad 1929; trans. Jerusalem 1971), 49.

2. Simon Baker, 'The hitherto impossible in photography is our speciality', *Air and Space* (Oct.–Nov. 1968).

3. Wohl, *The Spectacle of Flight*.

4. Albert Stevens, 'Exploring the stratosphere', *National Geographic* (Oct. 1934); William E. Kepner, 'The saga of Explorer I: man's pioneer attempts to reach space', *Aerospace Historian* (Sept. 1971); David Devorkin, *Race to*

the Stratosphere: Manned Scientific Ballooning in America (Washington DC, 1989).

5. Albert Stevens, 'Man's farthest aloft', *National Geographic* 69(1) (Jan. 1936), 59, 80; Albert Stevens, 'The scientific results of the world record stratospheric flight', *National Geographic* 69(5) (May 1936). The photograph was issued in a supplement to the May issue. It has not proved possible to trace a usable copy. This was not the first to show evidence of the curve of the Earth. On another flight, Captain Stevens had taken a photograph looking out over California from 23,000 feet to Mount Shasta, 331 miles away – the same distance as the horizon in the Explorer II photo. The horizon itself was much closer here, but because of the curve of the Earth just the top of the mountain was visible (NASM image 2005–35123). There may be earlier examples; anyone who has seen a ship's mast sinking behind the horizon has experienced a similar effect.

6. *The First Forty Years: A Pictorial Account of the Johns Hopkins Applied Physics Laboratory* (Baltimore, 1983), 13.

7. T. Bergstrahl, 'Photography from a V-2 rocket', Naval Research Laboratory (NRL) Progress Report, Washington DC, Aug. 1947; David Devorkin, *Science with a Vengeance: How the Military Created the US Space Sciences after World War II* (New York, 1992), chs 1–2.

8. Homer E. Newell, *High-Altitude Rocket Research* (New York, 1953), 89–103; Clyde Holliday, 'Preliminary report on high-altitude photography', *Photographic Engineering* I(1) (Jan. 1950); Clyde Holliday, 'Seeing the Earth from 50 miles up', *National Geographic* 98(4) (Oct. 1950).

9. The view from the V-2 was captured in the 1956 APL film *High Altitude Research* (NASM).

10. Devorkin, *Science with a Vengeance*, 144–6; Newell, *High-Altitude Rocket Research*, 283–4.

11. Bergstrahl, 'Photography from a V-2 rocket'; Devorkin, *Science with a Vengeance*, 145; Newell, *High-Altitude Rocket Research*, 283–8; *The First Forty Years*, ch. 2; 'So Columbus was right!', APL souvenir booklet, Oct. 1948. On the Columbus myth, see Russell, *Inventing the Flat Earth*. Columbus knew the Earth was round and expected to land in Asia; he rejected the idea that he had found a new continent. The myth was started in the early nineteenth century by Washington Irving.

12. In May 1954 the US Navy's Viking rocket took clear photographs of much of Mexico from a height of 158 miles (254 kilometres), R. C. Baumann and L. Winkler, *Photography from the Viking Rocket at Altitudes Ranging up to 158 Miles*, NRL report 4489 (Washington DC, 1955).

13. *Films from Space* (General Electric Company, 1959).

14. NASM X-15 file; *X-15: Research at the Edge of Space*, NASA EP-9 (n.d.).

15. Newhall, *Airborne Camera*, 113.

16. Devorkin, *Science with a Vengeance*, 146.

17. William R. Corliss, *Scientific Satellites*, NASA SP-133 (Washington DC, 1967), 717ff.; *Evening Star* (Washington), 28 Sept. 1959; *Washington Post*, 29 Sept. 1959; Explorer VI press conference transcript 28 Sept. 1959, NHO Explorer VI file 005819.

18. Henry G. Plaster, 'Snooping on space pictures', NHO (from a declassified CIA journal, fall 1964).

19. Burrows, *This New Ocean*, 303–4.
20. John Logsdon (ed.), *Exploring the Unknown*, NASA SP-4407 (1995), 156–61; Devorkin, *Science with a Vengeance*, 145; H. Wexler, 'Observing the weather from a satellite vehicle' (1954), in Logsdon, *Exploring the Unknown*.
21. TIROS files, NHO 00647-8; interview with William Stroud, Sept. 1973, NHO 002239. During the U-2 spy plane crisis, NASA released a photograph of Lake Baikal in the Soviet Union taken by TIROS on 4 April 1960, sending an implicit message to the Russians: 'Calm down – look what else we can do'. NASA photo release H 410 36–16.
22. *A Quasi-Global Presentation of Tiros III Radiation Data*, NASA SP-53 (1964), 11.
23. TIROS I press conference 22 Apr. 1960, NHO TIROS files 00647-8.
24. Letter from Homer E. Newell, 12 Aug. 1958, NHO file 005463.
25. Underwood's official title during the Apollo years was 'Manager, Operational Applications Office, Photographic Technology Division'. The following account is drawn from NHO Biographical Data Sheet; JSC OHA interview with Richard Underwood by Chick Bergen, 17 Oct. 2000; Watkins, *Apollo Moon Missions*; telephone interview and correspondence with the author, June–Sept. 2007; Richard Underwood, 'Lessons of the lenses', unattributed article in NHO file 006579; Richard Underwood, 'Aerial cameras, aerial films, and film processing', in *Earth Resources Survey Systems*, NASA SP-283 (1972); Schick and van Haften, *The View from Space*, 12–15.
26. 'Project Mercury', *Industrial Photography* (June 1961); 'A table and reference list documenting observations of the Earth', JSC-11710 (1976).
27. Newhall, *Airborne Camera*, 120–1.
28. Tom Wolfe, *The Right Stuff* (New York, 1980), 320–6; Burrows, *This New Ocean*, 284–92; Schick and van Haften, *The View from Space*, 11.
29. *Results of the Third US Manned Orbital Space Flight, October 3 1962*, NASA SP-12 (Washington DC, 1962); 'MA-8 post-launch Memo report 1: mission analysis', and 'Report 3: Air-ground voice and debriefing'; White, *The Overview Effect*, 27–30, 181–2, quoting from the Mercury team's *We Seven* (New York, 1962); Smith, *Moondust*, 282–4; Underwood interview, JSC, 5–7.
30. Schick and van Haften, *The View from Space*, 11–12; Pamela E. Mack, *Viewing the Earth* (Cambridge, MA, 1990), 39–40.
31. H. J. P. Arnold, 'The camera in space', *Spaceflight* (Dec. 1974), 442–53; P. D. Lowman, 'The earth from orbit', *National Geographic* (Nov. 1966), 669, quoted in Margaret Dreikhausen, *Aerial Perception* (Philadelphia, 1985), 44. Dreikhausen's study combines the findings of science with the insights of an artist.
32. H. J. P. Arnold, 'Gemini: the EVA photography', *Journal of the British Interplanetary* Society 37 (1984), 207–12; Francis French and Colin Burgess, *In the Shadow of the Moon* (Lincoln, NE, 2007), 29–33.
33. 'Lessons of the lenses'; Underwood interview, JSC OHA, p. 6.
34. Watkins, *Apollo Moon Missions*, ch. 3. My own experience of (literally) blowing the dust off old NASA technical publications in a deserted cul-de-sac of Manchester University library bears out Underwood's prophecy.

35. H. J. P. Arnold, 'Lunar surface photography: a study of Apollo 11', paper at 38th Congress of the International Aeronautical Federation, Oct. 1987, published by American Institute of Aeronautics and Astronautics.

36. Arnold, 'The camera in space'; Schick and van Haften, *The View from Space*, 11–12.

37. Underwood interview, JSC OHA, 11.

38. Schick and van Haften, *The View from Space*, 12; Underwood interview, JSC OHA 1–19.

39. Bruce K. Byers, *Destination Moon: A History of the Lunar Orbiter Program*, NASA TM X-3487 (Washington DC, 1977), 67–70; NHO Lunar Orbiter file 005 158; General Thomas S. Moorman, 'The ultimate high ground: space and our national security', 1998 Wernher von Braun memorial lecture.

40. Robert J. Helberg, *Lunar Orbiter Programme* (Seattle, WA, 1966), NHO file 005156; NASA news release 28 July 1966, NHO file 005158; Lunar Orbiter press kit 29 July 1966, NHO file 005159; Oran Nicks to Edgar Cortright, 13 June 1966, NHO file 005154.

41. Personal communication.

42. 'Transcript of discussion between Oran W. Nicks, Benjamin Milwitzky, and Lee R. Scherer of NASA and members of the National Academy of Public Administration, Washington, D.C. 12 Sept. 1968', NHO 001580, 106–9; Byers, *Destination Moon*, 225–7, 241–3.

43. 'Transcript of discussion, 12 Sept. 1968', 110–12.

44. NHO Lunar Orbiter file 005155; *Lunar Orbiter I Photographic Mission Summary*, NASA CR-782 (Washington DC, Apr. 1967); NASA release 66–251, 14 Sept. 1966, 'Lunar Orbiter circles Moon, snaps pictures'.

45. Cortright, *Apollo Expeditions to the Moon*, 78. Other NASA publications did choose the more dramatic landscape view of the picture, but always stressed the unfamiliar perspective.

46. NASA press photo 66-H-1146, as released 14 Sept. 1966.

47. Photo 67-H-218, re-enhanced Oct. 1966, released 2 Mar. 1967, and reproduced in *Lunar Orbiter I Preliminary Results*, NASA SP-197 (Washington DC, 1969), 88.

48. Joseph Karth biographical file, NHO file 001159; and see for example Karth, 'Technology of social progress', speech to 6th Goddard Memorial symposium, Mar. 1968, NASM file OS-505368-01; and see chapter 8 below.

49. Edgar Cortright (ed.), *Exploring Space with a Camera*, NASA SP-168 (Washington DC, 1968), 84–5; Newhall, *Airborne Camera*, 118; Arthur C. Clarke, *The Promise of Space* (New York, 1968), 149 and plate 31.

50. NASA press release 66–228, 23 Aug. 1966, 'Lunar Orbiter earth-from-moon photo attempted', NHO Lunar Orbiter file 005155; Oran W. Nicks, circulars to Lunar Orbiter progamme staff, Sept. 1966 and 22 Aug. 1967, NHO Lunar Orbiter file 005158; NASA, *The Lunar Orbiter: A Radio-Controlled Camera*, NASA Langley Research Center pamphlet (*c*.1969), inside back cover; NASA press release 68-23, 31 Jan. 1968, NHO Lunar Orbiter file 005158; L. J. Kosofsky and Farouk El-Baz, *The Moon as Viewed by Lunar Orbiter*, NASA SP-200 (Washington DC, 1970); David E. Bowker and J. K. Hughes (eds), *Lunar Orbiter Photographic Atlas of the Moon*, NASA SP-206 (Washington DC, 1971); NASA leaflet 'The eyes of Lunar Orbiter', NHO Lunar Orbiter file 005158; Cortright, *Exploring Space with a Camera*, 84–5.

51. NASA, *The Lunar Orbiter.*
52. Memo, 7 Nov. 1966, 'Earth photography from Lunar Orbiter missions', NHO Lunar Orbiter file 005156.
53. Transcript of discussion, 12 Sept. 1968, 115; Lunar Orbiter post-launch reports, 3–8 August 1967.
54. NASA release 67-14, 27 Jan. 1967, NHO Lunar Orbiter file 005158.
55. Richard Underwood, in Watkins, *Apollo Moon Missions*, 32; *Surveyor Program Results*, NASA SP-184 (Washington DC, 1969), 119–24.
56. *Evening Star* (Washington), 11 Jan. 1968.
57. *Surveyor Program Results*, 116–19; NHO Surveyor 7 files 5443–5, including Surveyor 7 press kit 4 Jan. 1968, Surveyor 7 mission status reports, Surveyor VII news conference 25 Feb. 1968, and press cuttings.
58. Martin Collins (ed.), *After Sputnik: Fifty Years of the Space Age* (Washington DC, 2007), 140–1. Surveyor 3's camera is now in the National Air and Space Museum, Washington DC.

Chapter 5 Blue marble

1. Applied Physics Laboratory (APL) DODGE press information, 1 July 1967, in NASM; *Washington Post*, 14 Oct. 1967; Kenneth F. Weaver, 'Historic color portrait of earth from space', *National Geographic* (Nov. 1967).
2. *Missiles and Rockets*, 7 Mar. 1966, 22.
3. NHO ATS files 005630–3; NASM ATS files 10639–667. The individual items listed below are from these files, mostly NHO.
4. *Evening Star* (Washington), 14 Dec. 1966.
5. *Aviation Week and Space Technology*, 19 Dec. 1966; *NASA News*, 2 Dec. 1976; NHO ATS-III files 005637–9, and Verner Suomi biographical file. ATS-III's colour camera used electronic rather than mechanical scanning, but at 4.5 miles its resolution of the Earth was less than half that of ATS-I.
6. NHO Verner Suomi biographical file 002248. A colour copy of this photograph and of others which could not be reproduced in this book can be found at the Earthrise website, www.earthrise.org.uk.
7. *Washington Post*, 11 Dec. 1966.
8. *North American Aviation* magazine Apollo 4 preview, NHO file 007233; *Baltimore Sun*, 10 Nov. 1967; *New York Times*, 10 Nov. 1967; NHO file 007233.
9. *Evening Star* (Washington), 12 Nov. 1967; *Life*, 24 Nov. 1967; *The Last Whole Earth Catalog* (New York, 1971). On Stewart Brand and the *Whole Earth Catalog*, see chapter 8.
10. British sources mention tortoises, American sources turtles. Siddiqi, *Challenge to Apollo*, 653–6; Turnill, *The Moonlandings*, 132–4; Scott and Leonov, *Two Sides of the Moon*, 227; www.astronautix.com.
11. Scott and Leonov, *Two Sides of the Moon*, 225. Technically this and the other Zond photographs, like those of Lunar Orbiter, showed Earth setting rather than rising.
12. Richard Underwood interview, JSC OHA; Siddiqi, *Challenge to Apollo*, ch. 15. Good information on the Soviet space programme can be found on the *Encyclopedia Astronautica* website at www.astronautix.com, and a

collection of Soviet Moon images at Don P. Mitchell's website, www. mentallandscape.com/C_CatalogMoon.htm.

13. Michael Collins, at an MSC press conference, quoted in Arnold, 'Lunar surface photography', 7; Richard Underwood to JSC History Office, 23 Dec. 1998 (JSC collection).

14. Edwin E. Aldrin (with Wayne Warga), *Return to Earth* (New York, 1973), 180; *Apollo 11 Mission Report*, Nov. 1969, including extracts from mission plan 79–80, in NHO file 005463.

15. Collins, *Carrying the Fire*, 349; Aldrin, *Return to Earth*, 236. Aldrin did have the camera for twenty-five minutes, however, and took one indistinct and distant photograph of Armstrong, back to the camera, working by the leg of the lunar module.

16. Arnold, 'Lunar surface photography', 1–2; Underwood interview, JSC OHA, 20; *Apollo 11 Mission Report*, Nov. 1969, NHO file 005463.

17. Richard Underwood interview, JSC OHA 36–8; Collins, *After Sputnik*, 151.

18. George Low, Deputy Administrator, to Dale Myers, Associate Administrator for Manned Space Flight, 28 Jan. 1972, and Myers' reply, 18 Feb. 1972, NHO 'Photography' file.

19. Watkins, *Apollo Moon Missions*, 31; Underwood to JSC, 23 Dec. 1998; Richard Underwood interview, JSC OHA 51–2.

20. Profile of Harrison Schmitt, *Washington Post* magazine, 30 Sept. 1982; Gene Cernan, *Last Man on the Moon* (New York, 1999), 324.

21. Richard Underwood interview, JSC OHA 51–2; *Aviation Week and Space Technology*, 15 Jan. 1973; NASA, *Apollo 17 Mission Report* (1973), 11–15.

22. *Congressional Record*, 22 Jan. 1973; *Miami Herald*, 1 Dec. 1972; *Kansas City Star*, 7 Dec. 1972.

23. *Chicago Sun-Times*, 21 Dec. 1972; *New Yorker*, 30 Dec. 1972; *Wall Street Journal*, 6 Dec. 1972 (editorial).

24. *Aerospace Perspectives* (Mar. 1973).

Chapter 6 An astronaut's view of Earth

1. The comment is widely quoted by Soviet cosmonauts and writers, but I have not been able to identify the original source for it.

2. Stanley G. Rosen, 'Space consciousness: the astronauts' testimony', *Michigan Quarterly Review* 18(2) (spring 1979), 279–99, at 288–91. For other collections of astronaut testimonies, see Ronald Weber, *Literary Responses to Space Exploration* (Athens, OH, 1985); White, *The Overview Effect*; Kevin W. Kelley (ed. for Association of Space Explorers), *Home Planet* (1988; London, 1991); Colin Fries, 'The green hills of earth: views of earth as described by astronauts and cosmonauts', *Quest* 10(3) (2003).

3. Rosen, 'Space consciousness', 288, 291; White, *The Overview Effect*, 15–23.

4. James Irwin with William A. Emerson, *To Rule the Night* (Philadelphia, 1973), 18–19; Schweickart, interviewed in Stewart Brand (ed.), *Space Colonies* (London, 1977), 34.

5. White, *The Overview Effect*, 20; Oleg Makarov, Preface to Kelley, *Home Planet*.

6. Collins, *Carrying the Fire*, 471; Michael Collins, 'Our planet: fragile gem in the universe', *Birmingham Post-Herald*, 1 Mar. 1972.

7. Cernan, *Last Man on the Moon*, 206–7; Irwin, *To Rule the Night*, 17; Aldrin, *Return to Earth*, 222.
8. Irwin, *To Rule the Night*, 60; Aldrin, *Return to Earth*, 236.
9. White, *The Overview Effect*, 182–7, 187–90; Cernan, *Last Man on the Moon*, 208–9; Kelley, *Home Planet*, caption 53; Irwin, *To Rule the Night*, 17; Noel, 'Re-entry', 170–1; Aldrin, *Return to Earth*, 239.
10. White, *The Overview Effect*, 15–23, 33; Weber, *Literary Responses*, 47–8, 52, 56–8; Daniel Greene, 'What next after you've walked on the Moon?', *National Observer*, 17 May 1975.
11. Schweickart, 'No frames, no boundaries', 5–6.
12. Henry S. F. Cooper, *A House in Space* (1976; London, 1977), 166–7; Irwin, *To Rule the Night*, 17–19.
13. William E. Honan of *Esquire*, July 1969, quoted in Weber, *Literary Responses*, 59; Scott and Leonov, *Two Sides of the Moon*, 289.
14. Cernan, *Last Man on the Moon*, 347; White, *The Overview Effect*, 128; Irwin, *To Rule the Night*, 17; Collins, *Carrying the Fire*, 473–4, 408–9.
15. Scott and Leonov, *Two Sides to the Moon*, 230–8; Schweickart quoted in *Philadelphia Enquirer*, 26 June 1980.
16. Schweickart, 'No frames, no boundaries', in Michael Katz, *Earth's Answer* (New York, 1977) 2–13. An edited version can be found in *Rediscovering the North American Vision,* issue 3 of *In Context* (summer 1983).
17. Scott and Leonov, *Two Sides to the Moon*, 236–8. The information about the music was supplied by Schweickart in an e-mail to NASM in 2002: see NHO file 001952.
18. White, *The Overview Effect*, 43, 250–6; Kelley, *Home Planet*, at picture 82, and Preface by Oleg Makarov; Scott and Leonov, *Two Sides to the Moon,* 310; Frank Borman in *Life*, 17 Jan. 1969.
19. Collins, *Carrying the Fire*, 470; White, *The Overview Effect*, 47.
20. Irwin, *To Rule the Night*, 60, 17–19; Smith, *Moondust*, 331.
21. Cernan, *Last Man on the Moon* 208–9, 322–4, 336–9.
22. White, *The Overview Effect*, 20; Smith, *Moondust*, 197.
23. Rosen, 'Space consciousness', 288–91. See also Smith, *Moondust*, ch. 2 and 69–70; Kelley, *Home Planet*, at pictures 138–9; and see below, chapter 8.
24. Rosen, 'Space consciousness', 294–5; Harold Masursky, G. W. Colton and Farouk El-Baz (eds), *Apollo over the Moon*, NASA SP–362 (1978), 254–5. See also Alfred Worden, *Hello Earth* (Los Angeles, 1974).
25. Irwin, *To Rule the Night*,18–19; Schweickart, 'No frames, no boundaries', 12.
26. *Washington Post*, 19 July 1966.
27. Cernan, *Last Man on the Moon*, 284–5, 339, 347; Collins, 'Our planet'. Al Bean talks about his paintings, including 'Mother Earth', in the extras to the DVD *Apollo 8: Leaving the Cradle* (Fox, 2003).
28. Smith, *Moondust*, 331, 256; Scott and Leonov, *Two Sides to the Moon*, 307–9. The episode is explored fictionally in Julian Barnes, *A History of the World in 10$^{1}/_{2}$ Chapters* (London, 1989), part 9.
29. White, *The Overview Effect*, 38–9; Rosen, 'Space consciousness', 288–91. After leaving NASA, Mitchell persuaded his former colleague Charles Duke to agree to do similar experiments on Apollo 16, but (apologised Duke) he was always too tired: Smith, *Moondust*, 64, 264.
30. Cooper, *A House in Space*, 166–7; Collins, *Carrying the Fire*, 470, 472; White, *The Overview Effect*, 46.

31. Mitchell, quoted in Rosen, 'Space consciousness'; last sentence added in Russell, *Global Brain*, 32.
32. Schweickart, 'Earth'.
33. Apollo 8 Post-Flight Press Conference, JSC 078–11. Before the mission Borman had anticipated seeing the Earth from the Moon, but said that what he most looked forward to was 'stepping out on to the carrier after a successful mission'. See chapters 1 and 2 above.
34. James Lovell, interview for 1999 PBS Nova TV documentary *To the Moon* (transcript in JSC).
35. *Congressional Record*, 22 Jan. 1973; Schmitt, 'The new ocean of space', 327–9; Rosen, 'Space consciousness', 292–4; Smith, *Moondust*, 282.
36. *Washington Post* magazine, 30 Sept. 1982.
37. Kelley, *Home Planet*, at picture 80; Collins, *Carrying the Fire*, 473–4; Oriana Fallaci, *If the Sun Dies* (1965; London, 1967), 160; White, *The Overview Effect*, 23–4.
38. Scott and Leonov, *Two Sides to the Moon*, 379; Irwin, *To Rule the Night*, 20; Smith, *Moondust*, 331. Gene Krantz found that the same principle held true on the ground: as flight controller for Apollo 11 he was too busy to savour the Moon landing, but as observer on Apollo 8 he was able to take in the enormity of what was happening and was overtaken by waves of emotion: Krantz, *Failure is Not an Option*, 242, 293–4.
39. Scott and Leonov, *Two Sides to the Moon*, 379–80; *Washington Post*, 13 Dec. 1994 (reporting on Roosa's death).
40. Cernan, *Last Man on the Moon*, 208–9; Aldrin, *Return to Earth*, 258, 300, 338, 306; Kelley, *Home Planet*, at pictures 138–9.
41. Weber, *Literary Responses*, 56–8; Schweickart, 'No frames, no boundaries', 4; Cernan, *Last Man on the Moon*, 347.
42. Cooper, *A House in Space*, 144, 148–51, 164–8.
43. Ibid., 135–7, 148–9, 166.
44. Ibid., 167–8; *Wall St. Journal*, 25 Apr. 1983, quoted in Weber, *Literary Responses*, 117.
45. Alexei Leonov, press release for Association of Space Explorers Congress, 1985, Association of Space Explorers website at www.space-explorers.org.
46. Kelley, *Home Planet*. See also the ASE website, at www.space-explorers.org.

Chapter 7 From Cold War to open skies

1. Winter, *Dreams of Peace and Freedom*, 99–100.
2. Fae L. Korsmo, 'The birth of the International Geophysical Year', *The Leading Edge* 26, 10 (Oct. 2007); M.I. Glassner, 'The Frontiers of Earth – and of Political Geography: the Sea, Antarctica, and Outer Space', *Political Geography Quarterly* 10 (1991), 218–19.
3. J. McCannon, 'To storm the Arctic: Soviet polar exploration and public visions of nature in the USSR, 1932–1939', *Ecumene* 2 (1995).
4. Klaus Dodds, 'To photograph the Antarctic: British polar exploration and the Falkland Islands and Dependencies Aerial Survey Expedition (FIDASE)', *Ecumene* 3(1) (1996); Glassner, 'The Frontiers of Earth'; Dian Olson Belanger, *Deep Freeze* (Boulder, CO, 2006), 32–3, 275–6.

5. Edgar Cortright, *Space Exploration: Why and How*, NASA (1965); Mack, *Viewing the Earth*, ch. 1, 52–5.

6. Cortright, *Space Exploration*; Logsdon, *Exploring the Unknown*, Vol. 3: *Using Space*, NASA SP–4407 (1998), 226–50; *Washington Post*, 11 Jan. 1969.

7. *Congressional Record*, 9 Jan. 1968; Borman, *Countdown*, 223–5; *Evening Star* (Washington), 2–6 July 1969; NASM file 1995–0025, 'Apollo 8 and 11 notes and letters'; NHO biographical file on Simon Bourgin, LEK1/3/1; and see below, chapter 8.

8. Brand, *Space Colonies*, 138–45, 98–103; and see below, chapter 8.

9. Cosgrove, 'Contested global visions: One-World, Whole-Earth, and the Apollo space photographs' *Annals of the Association of American Geographers* 84(2) (1994), 275–6.

10. Walter A. McDougall, *The Heavens and the Earth: A Political History of the Space Age* (New York, 1985), 174; Gerard De Groot, *Dark Side of the Moon: The Magnificent Madness of the American Lunar Quest* (New York, 2006), 266.

11. Winter, Dreams of *Peace and Freedom*, 5; Gene Cernan interviewed for the film *In the Shadow of the Moon* (2007); Aldrin, *Return to Earth*, 140; Smith, *Moondust*, 66–7. In the same way, the physicist Freeman Dyson was uneasy at his work on developing nuclear weapons in the Second World War. He responded by designing a rocket propelled upwards by a series of atmospheric nuclear explosions, which he regarded as a kind of swords into ploughshares project. His son George went in the opposite direction, living in a treehouse and building canoes. By way of explanation, Freeman sent George a copy of Kurt Vonnegut's novel *Slaughterhouse Five*; eventually the two were reconciled. Brower, *The Starship and the Canoe*.

12. National Security Council NSC 5520, in John M. Logsdon (ed.), *Exploring the Unknown: Selected Documents in the History of the U.S. Civil Space Program I: Organizing for Exploration*, NASA SP-4407 (Washington DC, 1995), 308–14; Dwayne A. Day, 'Cover stories and hidden agendas: early American space and national security policy', in R. Launius, J. Logsdon and R. Smith (eds), *Reconsidering Sputnik* (London, 2000); Burrows, *This New Ocean*, 166–70.

13. Rip Bulkeley, 'The Sputniks and the IGY', in Launius, Logsdon and Smith, *Reconsidering Sputnik*.

14. John Logsdon, *The Decision to Go to the Moon: Project Apollo and the National Interest* (Boston, 1970), 17; R. Cargill Hall, 'Origins of US space policy: Eisenhower, open skies and freedom of space', in Logsdon, *Exploring the Unknown*, 228; UN General Assembly Resolution 13489 (xiii), 13 Dec. 1958, 'Question of the peaceful use of outer space'.

15. McCurdy, *Space and the American Imagination*, 74–5; National Aeronautics and Space Council, 'US policy on outer space', 26 Jan. 1960, reproduced in Logsdon, *Exploring the Unknown*, 362.

16. The results of satellite reconnaissance were later summed up by a senior military intelligence officer: 'Corona opened the Sino-Soviet bloc to scrutiny. It proved the missile gap non-existent, and therefore saved hundreds of millions, if not billions, of dollars that we would have spent on strategic programs. It pinpointed all of the Soviet missile sites and gave us accurate orders of battle for their forces. Finally, as the first national tech-

nical means of verification, it made arms control and START treaties possible'. Moorman, 'The ultimate high ground'.

17. Logsdon, *Decision to Go to the Moon*, ch. 1, 22–6 and 37–8; DeGroot, *Dark Side of the Moon*, ch. 8, 160–2; Anne M. Platoff, 'Where no flag has gone before: political and technical aspects of placing a flag on the moon', NASA Contractor Report 188251 (1993).

18. Myers S. McDougal, Harold D. Lasswell and Ivan A. Vlasic, *Law and Public Order in Space* (New Haven, CT, 1963), 1103. This massive treatise pleaded with statesmen to avoid repeating the 'tragic destiny of a divided Earth' in space, and instead to build 'a commonwealth of dignity for all advanced forms of life' (at 1025–6). A series of international agreements governing space made it obsolete almost as soon as it was published, although its 2,619 footnotes and numerous appendices ensured that it was not without value.

19. George Dyson, *Project Orion* (New York, 2001); UN Resolution 1884 (xviii), 17 Oct. 1963, and final resolution 1962 (xviii), 13 Dec. 1963; Carl Q. Christol, *The Modern International Law of Outer Space* (New York, 1982), 20–5. In 1962 the UN had made a Declaration of Principles on outer space, which formed the basis of the 1963 act.

20. J. E. S. Fawcett, *Outer Space: New Challenges to Law and Policy* (Oxford, 1984), 3–6.

21. Christol, *The Modern International Law of Outer Space,* ch. 2; Fawcett, *Outer Space*, 1–21. The treaty represented a slight step back from the 1962 Declaration of Principles, which proposed demilitarising all of outer space without distinction. On the question of limits, 'airspace' effectively went up to about 10 miles (16 kilometres), the limit of normal flight, while useful orbits began at about 75 miles (120 kilometres). Was there perhaps some intermediate zone where neither set of rules applied? The X-15 rocket planes clinging to the last vestiges of atmosphere could reach up to 67 miles (108 kilometres), and pilots flying above 50 miles were given astronaut's wings. The first Mercury 'space' flight reached 116 miles without going into orbit.

22. *New York Times*, 22 Dec. 1968; *Boston Globe*, 28 Dec. 1968; Bill Anders interview, JSC, *To the Moon*, PBS Nova TV documentary, 1999.

23. *Christian Science Monitor*, 7 Jan. 1969; *Chicago Sun-Times*, 29 Dec. 1968.

24. Paine Papers, LOC, box 43 (Apollo 8), 28 May 1969.

25. C. S. Lewis, 'Will we lose God in Outer Space?', *Christian Herald*, Apr. 1958, reprinted as 'Rockets and religion' in Lewis, *The World's Last Night* (New York, 1987). Celebi afterwards claimed to bring Jesus' greetings back to Earth from a height of some 300 metres: Frank Winter, 'Who first flew in a rocket?', *Journal of the British Interplanetary Society* (July 1992), quoted in Burrows, *This New Ocean*, 22.

26. James Lovelock, *The Ages of Gaia* (1988; 2nd edn, New York, 1995), 172–3.

27. *New York Times*, 24 Aug. 1962, in NHO file 006774, 'Impact of space: religion'.

28. *Washington Post*, 7 May 1962; *Evening Star* (Washington), 6 July 1963; undated cutting in NASA Current News file, *c.* 29 Dec. 1968.

29. Martin J. Heinecken, *God in the Space Age* (Philadelphia, 1959), 7; Scott and Leonov, *Two Sides to the Moon*, 209–10.

30. Moscow Domestic Service in Russian, 12 May 1962, reported in the bulletin *USSR International Affairs*; *Congressional Record*, 24 May 1961; NHO file 006774.

31. Fallaci, *If the Sun Dies*, 384–5; Wolfe, *The Right Stuff*.

32. DeGroot, *Dark Side of the Moon*, 107–8; *Catholic Standard*, 9 Mar. 1962; *Boston American Record*, 26 July 1962; NHO file 006774. Cooper's prayer seemed to lose its way slightly, concluding, 'Help us in our future space endeavors that we may show the world that a democracy really can compete, and is still able to do things in a big way, and still is able to conduct various scientific, very technical programmes in a completely peaceful environment.' The reality was sometimes less clean-cut. The Italian journalist Oriana Fallaci, whose memoir describes a friendship with the astronaut Pete Conrad, later sent him her grandmother's seventeenth-century silver cross to take up in his Mercury capsule, but he sent it back with an excuse: it would have had to have been disinfected first. Fallaci, *If the Sun Dies*, 394.

33. *Houston Post*, 17 Oct. 1967.

34. Wernher von Braun, 'Space travel and our technical revolution', paper at American Rocket Society, New York, 4 Apr. 1957, and Evangelical Academy, Locum, Germany, 28 Feb 1958, quoted in Heinecken, *God in the Space Age*, 7; Fallaci, *If the Sun Dies*, ch. 20. Von Braun later sent long and thoughtful replies to individuals who wrote to him on matters of faith, stressing that Christianity could evolve with science while remaining true to its core: NHO file 006774 (Oct.–Dec. 1971).

35. Heinecken, *God in the Space Age*, 10–12, 190–203, and ch. 4.

36. Wikipedia, 'Pope Pius XII', accessed 25 Jan. 2008, citing a speech of 1939; *Los Angeles Mirror*, 20 Sept. 1956; *New York Times*, 21 Sept. 1956; C. L. Sulzberger, 'The theology of race and space', *New York Times*, 15 Oct. 1962. The key issue was whether Christ's sacrifice had been for all life, or for the human race alone – in other words, whether God had made separate dispensations for other forms of life, or whether one plan covered all. C. S. Lewis, as ever, put it most clearly. The tendency of modern Catholic theology seems to have been towards the universalist view, which in turn implies missionary activity in space and the baptism of aliens. The debate is over four centuries old. John Wilkins, *The Discovery of a World in the Moon* (1638), 185–6; Cressy, 'Early modern space travel'; Stephen Dick, *Plurality of Worlds* (Cambridge, 1982); Dick, *The Biological Universe*; Stephen Dick (ed.), *Many Worlds: The New Universe, Extraterrestrial Life and the Theological Implications* (Philadelphia, 2000); Heinecken, *God in the Space Age*, 115–46; Lewis, 'Will we still love God in outer space?'

37. *Herald Tribune*, 11 Jan. 1965; James Webb to Pope, 27 Jan. 1965, NHO file 006774; *New York Times*, 2 May, 7 June, 11 June, 17 July 1965; *Evening Bulletin* (Philadelphia), 9 June 1965.

38. Thomas Paine to Rev. Luigi Raimondi, 16 Dec. 1968, Paine Papers, LOC, box 22; *Sunday Bulletin* (Philadelphia) 16 Feb. 1969; NHO file 006774; NASA, *Space Quotes*, June 1969; *Liturgical Arts* (Nov. 1967). On Borman's visit to Rome in 1969, see chapter 2 above.

39. When a San Francisco Baptist minister wrote to suggest sending a Christian representative into space, NASA politely deflected the enquiry by sending a copy of the criteria for astronaut recruits, 'without regard to sex, religion,

race, or national origin': Office of Public Affairs to Rev. Belton C. Currington, Good Shepherd Baptist Church, San Francisco, NHO file 006774. NASA did, however, make encouraging responses in 1969 to a proposal from Congressmen for a 'chapel of the astronauts' at the Manned Spaceflight Center: Thomas Paine to Joseph Karth, 16 May 1969, Joseph Karth biographical file NHO 001159.

40. Aldrin, *Return to Earth*, 45; Borman, *Countdown*, 194–5; Frank Borman interview, JSC OHA, 1999; and see chapter 2 above.

41. *Washington Post*, 22 Dec. 1968. The *Birmingham News* (Alabama) took the Protestant line that exact observance of individual feasts was not important (20 Dec. 1968).

42. Apollo 8 crew briefing, 7 Dec 68, and Apollo 8 postflight press conference, JSC 078–11; Apollo 8 group conference 1998, 26.

43. Thomas Paine to Staffan Wennberg, 12 Aug. 1968, Paine Papers, LOC, box 22; C. P. Snow, in *Look*, 4 Feb. 1969.

44. Borman, *Countdown*, 194–5. The story is pieced together, from interviews, by Robert Zimmerman in *Genesis*, ch. 10, and filled out here by Bourgin's notes in NASM file 1995–025, 'Apollo 8 and 11 notes and letters'.

45. Interviews with Glynn Lunney, 1999, and James Lovell, 1999, JSC OHA; Apollo 8 group conference, 1998; Turnhill, *The Moon Landings*, 168–71.

46. Borman, *Countdown*, 223–5; *Parade*, 23 Feb. 1969.

47. *Kansas City Star*, 26 Dec. 1968; *Washington Post*, 26 Dec. 1968; *Evening Star* (Washington), 1 Jan. 1969; *Miami Herald*, 29 Dec. 1968.

48. *Washington Post*, 28 Dec. 1968; Frank Borman biographical file, NHO 000212.

49. Memorandum by R. G. Rose, 29 Jan. 1969, in Apollo 8 archives, JSC 078–11. The prayer was that for 'Vision, faith and work' by G. F. Weld. There were at that period only three Episcopalian churches in the Houston area, all of them closely connected, and their members also included the chief mission controller Chris Kraft. For a conversation on this point I am indebted to Bill Larsen at JSC.

50. Frank Borman, 'Message to Earth', *Guideposts* magazine (New York), Apr. 1969, 1–7; Paine Papers, LOC, box 43; *Detroit News*, 27 Apr. 1972; NHO file 006774; *Chicago Tribune*, 8 Dec. 1969.

51. *New York Times*, 25 and 28 Dec. 1968; *Parade*, 23 Jan. 1969; *Los Angeles Times*, 29 Dec. 1968.

52. Bill Anders interview, JSC, *To the Moon*, PBS Nova TV documentary, 1999; Julian Scheer obituary, 4 Sept. 2001, at www.space.com; JSC file 078–11.

53. *New York Times*, 25 Dec. 1968.

54. Aldrin, *Return to Earth*, 232–3; Daniel M. Harland, *The First Men on the Moon* (New York, 2007), 252; James R. Hansen, *First Man: The Life of Neil Armstrong* (New York, 2005), 487–8.

Chapter 8 From Spaceship Earth to Mother Earth

1. Hannah Arendt, *The Human Condition* (1958; Chicago, 1963), prologue 1–6; Hannah Arendt, 'Man's conquest of space', *American Scholar* 32 (autumn 1963), in Patrick Gleeson (ed.), *America, Changing* (Columbus, OH, 1968), 418–29.

2. William F. Buckley, 'Flat-earth liberals', *National Review*, 29 July 1969; McLuhan, quoted in Yakov Gaarb, 'The use and abuse of the whole earth image', *Whole Earth Review* 45 (Mar. 1985), 19–20; Anthony Lewis, 'Heroic materialism is not enough', *New York Times*, 20 July 1969, quoted in Michael L. Smith, 'Selling the moon', in R. W. Fox and T. J. J. Lears (eds), *The Culture of Consumption* (New York, 1983), 207; Kurt Vonnegut, 'Excelsior! We're going to the Moon! Excelsior!', *New York Times* magazine, 13 July 1969, 9–11, quoted in Smith, 'Selling the moon', 207; Amitai Etzioni, *The Moon-Doggle* (New York, 1964), 197–8.

3. William Irwin Thompson, *The Edge of History* (1971; New York, 1990), 142–3.

4. Quoted in Cosgrove, 'Contested global visions', 280–1.

5. Wells's story was well known to the quantum physicist Werner Heisenberg, who took it as a cautionary tale and was prompted by it to assign the patent on the nuclear chain reaction to the British Admiralty for safe keeping: Lovat Dickson, *H. G. Wells and his Turbulent Times* (London, 1972), 269, quoted in William Irwin Thompson, *Passages about Earth: Explorations of the New Planetary Culture* (New York, 1973), 56–7.

6. W. W. Wagar, *H.G. Wells and the World State* (New Haven, CT, 1961), 58–66; Denis Livingston, 'Science fiction models of future world order systems', *International Organization* 25(2) (1971); J. D. Bernal, 'The world, the flesh and the devil', at www.marxists.org/archive/.

7. Tsiolkovsky, *Beyond the Planet Earth*, chs 11–12; Arthur C. Clarke, quoted in Bainbridge, *The Spaceflight Revolution*, 150. On this theme, see also chapter 3 above.

8. Fallaci, *If the Sun Dies*, 25–8. The preface explains that the book is a subjective diary of her encounters rather than an exact record. At the time, Bradbury rode a bicycle and didn't have a TV. After the interview, Bradbury's wife confided in her that in twenty years of marriage she had never heard her husband say such things: 'It kind of shocked me.' But she agreed with him.

9. Louis J. Halle, 'Why I'm for space exploration', *New Republic*, 6 Apr. 1968, quoted in Smith, 'Selling the moon', 207; Brand, *Space Colonies*, 22–31 and passim; Gerald O'Neill, *The High Frontier* (London, 1977), chs 2–3, also 38–9; Jesco von Puttkamer, 'Space: a matter of ethics – towards a new humanism', in Eugene M. Emme (ed.), *Science Fiction and Space Futures* (San Diego, CA, 1982), 209.

10. McDougall, *The Heavens and the Earth*, ch. 21; Bizony, *The Man Who Ran the Moon*.

11. Marshall McLuhan, *Understanding Media* (New York, 1964), 3, 343.

12. Joseph E. Karth, 'Potential of oceanography', speech to National Space Club (based in Washington DC), 18 Jan 1966.

13. Garrett Hardin, *Exploring New Ethics for Survival: The Voyage of the Spaceship Beagle* (New York, 1972), 22; Fred Turner, *From Counterculture to Cyberculture: Stewart Brand, the Whole Earth Network, and the Rise of Digital Utopianism* (Chicago, 2006), 50–8; Andrew Kirk, *Counterculture Green* (Lawrence, KS, 2007), 56–62.

14. Buckminster Fuller, 'Vertical is to live – horizontal is to die', *American Scholar* 39 (winter 1969); Buckminster Fuller, *Operating Manual for Spaceship Earth* (Carbondale, IL, 1969); Newhall, *Airborne Camera*, 118; Brand, *Space Colonies*, 55.

15. Fuller, 'Vertical is to live', 47; Hal Aigner, 'Buckminster Fuller's World Game', *Mother Earth News*, Dec. 1970, 62–8.

16. Turner, *From Counterculture to Cyberculture*, ch. 2; *Mother Earth News*, Dec. 1970. Brand's other priorities were Buckminster Fuller's 'World Game', contingency planning for environmental disasters, and a 'wet NASA' to research and protect the oceans, as with Antarctica, as a fragile, transnational environment.

17. Steve Jobs, Stanford University commencement speech, 12 June 2005, quoted at Wikipedia under 'Whole Earth Catalog' (accessed Dec. 2007).

18. A close relative was *Mother Earth News* ('it tells you how'), an underground magazine published in Ohio 'to present the HOW of alternative life styles', 'edited by, and expressly for, today's influential "hip" young adults. The creative people. The doers. The ones who make it all happen.' It didn't share the *Whole Earth Catalog*'s fascination with space but still regarded Brand as 'a giant'. *Mother Earth News*, Jan. 1970, May 1970; Kirk, *Counterculture Green*, 85–6.

19. Kirk, *Counterculture Green*, 1–2.

20. Brand has left two accounts: 'The first whole earth photograph', in Katz, Marsh and Thompson, *Earth's Answer*, 184–8 (papers given at the 1974 and 1975 Lindisfarne conferences on the theme of 'exploring planetary culture'); and 'Whole earth origin', in the Rolling Stone book *The Sixties*, reproduced on Brand's website, http://sb.longnow.org/Home. Turner, *From Cyberculture to Counterculture*, plate 4 shows Brand in October 1966 wearing one of his whole Earth campaign buttons.

21. *San Francisco Chronicle*, 22 Mar. 1966.

22. Personal communication, Dec. 2007.

23. The film is NASM FA 00840, *ATS III: The First Color Movie of Planet Earth* (Nov. 1967). It was produced by Byron motion pictures of Washington DC, NASA's film printer, so the steep $49 price should, under NASA's standard 'no profit' contract, have reflected the company's costs.

24. 'Washington goes to the Moon', transcript of a NASA film, JSC; comments by Howard McCurdy and Roger Launius.

25. John McConnell, 'The history of the Earth flag', *Flag Bulletin*, Mar.–Apr. 1982 (Flag Research Center) and at www.earthsite.org; information from John and Anna McConnell. There was one glitch, however: 'In the rush to make the first flags the colors in the screening of Earth were reversed . . . the ocean is white and the clouds are blue.'

26. Copies of the Earth Day badges and flags are in the collection of the Smithsonian National Museum of American History at 1997.0355.

27. National Museum of American History at 1992.3134.

28. Thomas R. Huffman, 'Defining the origins of environmentalism in Wisconsin: a study in politics and culture', *Environmental History Review* (fall 1992); Marc Mowrey, *Not in Our Backyard: The People and Events that Shaped America's Modern Environmental Movement* (New York, 1993).

29. Gaylord Nelson, 'History of Earth Day', in the 'Earth Day' file at the Smithsonian National Museum of American History; Worster, *Nature's Economy*, 356–8.

30. Marcy Darnovsky, 'Stories less told: histories of US environmentalism', *Socialist Review* (Oct. 1992); *New York Times*, 23 Apr. 1970; *Life*, 24 Apr. 1970.

31. Barbara Ward, *Spaceship Earth* (New York, 1966), 17–18; Unesco, *Use and Conservation of the Biosphere* (Paris, 1970); Peder Anker, 'The ecological colonisation of space', *Environmental History* 10(2) (Apr. 2005), 6–8; Worster, *Nature's Economy*, 369–70; Joel B. Hagen, *An Entangled Bank: The Origins of Ecosystem Ecology* (New Brunswick, NJ, 1992), 189–97. On Adlai Stevenson, see chapter 3 above.

32. Barbara Ward and René Dubos, *Only One Earth: The Care and Maintenance of a Small Planet* (London, 1972), 9, 261, xvii; Dubos, 'A theology of the earth', in *A God Within*. Dubos's ideas anticipated the Gaia hypothesis: see chapter 9 below.

33. R. C. Mitchell et. al., 'Twenty years of environmental mobilization', in Riley E. Dunlap and Angela G. Mertig (eds), *American Environmentalism* (New York, 1992), 11–26.

34. Charles T. Rubin, *The Green Crusade* (New York, 1994), 6–7; Anne Chisholm, *Philosophers of the Earth: Conversations with Ecologists* (London, 1972).

35. Kenneth Boulding, 'The economics of the coming Spaceship Earth', paper given in Washington DC, Mar. 1966, reprinted in his *Beyond Economics* (Michigan, 1968), 275–87. Elsewhere, Boulding compared Tokugawa Japan to 'a spaceship economy . . . developing in a chrysalis. In 1868 they came out into the modern world and just flew': Chisholm, *Philosophers of the Earth*, 25–38.

36. Donna Meadows et al. (eds), *The Limits to Growth* (New York, 1972); *Mankind at the Turning Point: The Second Report to the Club of Rome* (New York, 1974), quoting A. Gregg, 'A medical aspect of the population problem', *Science* 121 (1955), 681.

37. James Cornell and John Surowiecki, *The Pulse of the Planet: A State of the Earth Report from the Smithsonian Institution Center for Short-Lived Phenomena* (New York, 1972), ix–xi. The Center itself seems to have been a short-lived phenomenon.

38. Thompson, 'The deeper meaning of Apollo 17', and its revision as chapter 1 of his *Passages about Earth*.

39. Thompson, *Passages about Earth*, 187–93 and 144–5; William Irwin Thompson, 'Sixteen years of the New Age', in W. Thompson and D. Springler, *Reimagination of the World* (Santa Fe, 1991), 5–6. Thompson acknowledged the inspiration of the mystic Teilhard de Chardin, the astronaut Edgar Mitchell and the novelist Doris Lessing.

40. Turner, *From Counterculture to Cyberculture*, 122–4; Kirk, *Counterculture Green*, 164–5 and ch. 5 generally.

41. Brand, *Space Colonies*, 98–103, and see chapter 7 above. Schweickart then introduced Brand to Jesco von Puttkamer, an astrofuturist former member of Wernher von Braun's circle now in charge of coming up with long-range ideas for NASA and, like Brand, an advocate of space colonies.

42. Schweickart, 'No frames no boundaries'; Brand, *Space Colonies*, 110–14, 138–45.

43. Turner, *From Counterculture to Cyberculture*, 126–8; Kilgore, *Astrofuturism*, ch. 5.

44. Brand, *Space Colonies*, 44, 51, 53.

45. Ibid., 146–8; *San Francisco Chronicle*, 11 and 14 Aug. 1977.

46. Kirk, *Counterculture Green*, 176–81.

47. UPI report 15 Apr. 1979, Russell Schweickart biographical file, NHO; *Washington Post*, 11 July 1980; *Washington Post* magazine, 30 Sept. 1982; White, *The Overview Effect*, 190–1.

48. *No Frames, No Boundaries* (1982), LOC; White, *The Overview Effect*, 69–70.

49. Jonathan Schell, *The Fate of the Earth* (New York, 1982), 153–4. Schell was aware (pp. 77–8) of Lewis Thomas's analogy of the Earth as a living cell – Lewis Thomas, *The Lives of a Cell* (New York, 1974) – but not apparently of the recent Gaia hypothesis.

50. Carl Sagan et al., 'Global atmospheric consequences of nuclear war' (1983) at http://ntrs.nasa.gov/archive/nasa/casi.ntrs.nasa.gov/19900067303 _1990067303.pdf; Paul Ehrlich et al., *The Cold and the Dark: The World after Nuclear War* (New York, 1984).

51. E. P. Thompson, *The Sykaos Papers* (London, 1988), 454. Thompson's novel was a spirited riposte to the lofty 'space fiction' of Doris Lessing, a fellow former communist and supporter of nuclear disarmament. While Thompson wrote a rumbustuous pamphlet called *Protest and Survive* denouncing civil defence as a con trick, Lessing advocated private nuclear shelters as a means of human survival.

52. Worster, *Nature's Economy*, 387.

53. Donna Hathaway, 'The promises of monsters', in L. Grossberg et al. (eds), *Cultural Studies* (London, 1992), 317–19. The action itself never happened; it remained a subject for cultural studies.

54. Yakov Gaarb, 'The use and abuse of the whole earth image', *Whole Earth Review* 45 (Mar. 1985), and 'Perspective or escape? Ecofeminist musings on contemporary earth imagery', in Irene Diamond and Gloria Bernstein (eds), *Reweaving the World* (San Francisco, 1990), 264–78, 305–8.

55. Hathaway, 'The promises of monsters', 317–19; Ann Oakley, *Gender on Planet Earth* (Oxford, 2002), 135.

Chapter 9 Gaia

1. Thomas, *The Lives of a Cell*, quoted in the epigraph to Lovelock, *The Ages of Gaia*; James Lovelock, *Gaia: A New Look at Life on Earth* (1979; Oxford, 1987), Introduction.

2. Worster, *Nature's Economy*, 378–87; Lovelock, *Gaia*, ix–x, 152.

3. James Lovelock, *Homage to Gaia: The Life of an Independent Scientist* (Oxford, 2000), 227; (sic: Lovelock may have meant 'the crescent').

4. Lovelock, *Gaia*, 3–4; Hagen, *An Entangled Bank*, 189–97; Dubos, 'A theology of the earth'.

5. Loren Eiseley, *The Firmament of Time* (1960; London, 1961), ch. 1 (based on lectures given in 1959); Worster, *Nature's Economy*, 387.

6. Liberty Hyde Bailey, *The Holy Earth* (1915; New York, 1980), 5–15.

7. Vladimir I. Vernadsky, *The Biosphere* (1926; New York, 1998), 41, 43–5 and Lynn Margulis et al., 'Foreword', 14–19; E. Suess, *The Face of the Earth*, i (Oxford, 1904), 1.

8. Lovelock, *The Ages of Gaia*, 192–3.

9. Ibid., 221; Lovelock, *Homage to Gaia*, 237–9.

10. Lynn Margulis, *The Symbiotic Planet* (London, 1998), 147–8; Lovelock, *Homage to Gaia*, 233–41; Lovelock, *Gaia*, ch. 1; James Lovelock and Lynn

Margulis, 'The Gaia hypothesis', *CoEvolution Quarterly* 6 (summer 1975); Lynn Margulis (ed.), *Proceedings of the Second Conference on the Origins of Life* (Washington DC, 1971), 39–40, 43–4, 171, 201, 205–7. See also Jeanne McDermott, 'Lynn Margulis', *CoEvolution Quarterly* 25 (spring 1980).

11. On current population projections, that point will be reached about 2050. Lovelock, *Gaia*, 12, 132, 40–2; Lynn Margulis to *The Ecologist* 16(1) (1986), 52–3; Howard Odum, *Environment, Power and Society* (1971), quoted in Anker, 'Ecological colonisation of space', 6–8.

12. Lovelock, *Gaia*, xiii, 148–9; *The Ages of Gaia*, xx–xxi.

13. Lovelock, *Homage to Gaia*, 345 and ch. 7; *Gaia*, 123 and Preface; James Lovelock, *The Revenge of Gaia* (London, 2006); John Leslie, *The End of the World* (London, 1996), 48–9.

14. Lovelock, *The Ages of Gaia*, 212; *Homage to Gaia*, xi–xix, 3. Despite this, Smith took the idea seriously.

15. Lovelock, *The Ages of Gaia*, 60–1, 19; for Arendt, see the opening of chapter 8 above.

16. Jon Turney, *Lovelock and Gaia* (London, 2003); Lovelock, *The Revenge of Gaia*, 31–2.

17. Margulis, *The Symbiotic Planet*, ch. 8.

18. Lovelock, *The Revenge of Gaia*, 2.

19. Lovelock, *Homage to Gaia*, 316–19; Lovelock, *The Ages of Gaia*, ch. 9; Hugh Montefiore, *The Probability of God* (London, 1985).

20. Richard Underwood, 'Lessons of the lenses'.

21. White, *The Overview Effect*; Schick and van Haften, *The View from Space*.

22. See Association of Space Explorers website at www.space-explorers.org.

23. Darnovsky, 'Stories less told', 36–40; *American Demographics* (Apr. 1990), 40–1; NASA website at www.nasa.gov.

24. Quoted in James Baker, *Planet Earth: The View from Space* (Cambridge, MA, 1990), v.

25. Lovelock, *Homage to Gaia*, xvii–xviii.

26. Winter, *Dreams of Peace and Freedom*, 177–82.

27. Steven Weinberg, *The First Three Minutes* (1977; Glasgow, 1978), 148–9.

28. Jon Turney, 'Telling the facts of life: cosmology and the epic of evolution', *Science as Culture* 10(2) (2001), 239; Carl Sagan, *Pale Blue Dot: A Vision of the Human Future in Space* (New York, 1994), 174–5 and ch. 1.

29. Sagan, *Pale Blue Dot*, 1–7; Sagan et al., *Murmurs of Earth* (New York, 1978).

30. Connie Barlow, *Green Space, Green Time* (New York, 1997), 11–12, 150–8, 212, 236–7, and ch. 5.

31. Al Gore, *Earth in the Balance* (London, 1992), 384–5; *Washington Post*, 3 Mar. 1998; Schweickart biographical file, NHO.

32. Al Gore, *An Inconvenient Truth* (London, 2006).

33. Donald Goldsmith (ed.), *The Quest for Extraterrestrial Life* (Mill Valley, CA, 1980), 2, vi.

34. Dick, *The Biological Universe*; Frank Drake and Dava Sobel, *Is Anyone Out There?* (New York, 1991).

35. Fallaci, *If the Sun Dies*, 203–4; Drake and Sobel, *Is Anyone Out There?*, Preface; McDougal, Lasswell and Vlasic, *Law and Public Order in Space*, 1025–6. (see above, p. 217, n.18).

36. This account draws largely on Dick, *The Biological Universe*, ch. 8.

37. James Lawrence Powell, *Mysteries of Terra Firma: the Age and Evolution of the Earth* (New York, 2001); Peter Ward and Donald Brownlee, *Rare Earth* (New York, 2000).

38. Ward and Brownlee, *Rare Earth*, 57, 61, 282–3.

Chapter 10 The discovery of the Earth

1. Quoted in Smith, *Moondust*, 57.

2. Patrick Moore, Preface to *The Race into Space: Man's First 50 Steps into the Universe*, Brooke Bond Oxo (1969); Schmitt, 'The new ocean of space'.

3. White, *The Overview Effect*, 65–7, 4–5; Wyn Wachorst, *The Dream of Spaceflight* (New York, 2000), 92–5.

4. Sagan, *Pale Blue Dot*, 1–7, 331–4.

5. Charles P. Boyle, *Space among Us: Some Effects of Space Research on Society* (Washington DC, 1974), quoted in Weber, *Literary Responses*, 71–2, 81n; Weber, *Literary Responses*, ch. 4, 61, 121; Symposium on the Value of Space Exploration, National Geographic Society, Washington DC, 1994, session 3, 'Cultural value of space exploration'; Robert Phillips, *Moonstruck: An Anthology of Lunar Poetry* (New York, 1974).

6. Montgomery, *The Moon and the Western Imagination*, 40; Wachorst, *The Dream of Spaceflight*, 93.

7. See chapter 3 above.

8. Thomas, *The Lives of a Cell*, quoted at the start of Lovelock, *Ages of Gaia*.

9. Sagan, *Pale Blue Dot*, 3.

10. Collins, *Carrying the Fire*, 471.

11. The pictures, including a real-time video of Earthrise, can be seen at the Japanese Agency website, http://www.selene.jaxa.jp.

12. *Observer*, 20 Jan. 2008, review section p. 3; Cosgrove, 'Contested global visions'.

13. Corfield, *Time and the Shape of History*, 215–16, 284 n 58.

14. Thomas Kuhn, *The Structure of Scientific Revolutions* (revised edn, Chicago, 1970).

Index